MONITORING OF RESPIRATION AND CIRCULATION

J.A. BLOM

CRC PRESS

Boca Raton London New York Washington, D.C.

Library of Congress Cataloging-in-Publication Data

Catalog record is available from the Library of Congress

Direct all inquiries to CRC Press LLC, 2000 N.W. Corporate Blvd., Boca Raton, Florida 33431.

Visit the CRC Press Web site at www.crcpress.com

International Standard Book Number 0-8493-2083-6
Printed in the United States of America 1 2 3 4 5 6 7 8 9 0
Printed on acid-free paper

Foreword

"Although the word *monitoring* is often thought to refer to the act of obtaining a measurement, the proper use of the term in anesthesia is to observe *and* control. That is, monitoring is both the obtaining and the use of information."[1] This text is about monitoring the important variables that are indicators of how well a patient's respiratory and circulatory systems function.

Measurements are extremely important in medicine. They form the basis of every diagnosis and hence of every therapy. Collecting information about the patient's health state is the central issue in the medical diagnostic process. Much of that information is collected through asking questions and using the senses (vision, audition and proprioception). The color of the skin and its stiffness, the movements of the chest and of the pulse at the wrist, the sounds of heart and lungs, the apparent humidity of the eyes and the size of the pupils are all important information sources that are readily available. This type of information, which is usually not quantitative and the interpretation of which requires much medical knowledge and experience, is not treated in this text. The text is limited to those measurements for which technology can provide a significant contribution and that would be impossible without advanced medical technical instrumentation. The text is moreover limited to measurements of physiological variables, i.e., those variables that can be directly measured, such as blood pressure and heart rate. It touches only briefly on electrical variables; the electrocardiogram (ECG) is mentioned, but not, for instance, the electroencephalogram (EEG) or the electromyogram (EMG). Also, only one-dimensional signals are considered, although we recognize that two-, three- and four-dimensional "images" are increasingly important sources of information.

If we want to extract information from measurements, we need to know what the measurements mean. The meaning is provided by a model. The model describes the important notions, which must be extracted from the measurements, such as a "heart rate" from an arterial blood pressure signal. It is also important to know the approximate reliability or accuracy of the extracted parameters. It is these parameters, after all, that describe a patient's health status and that are used to arrive at a therapy.

We can distinguish between several important classes of physiological variables:

- electric potentials (ECG, EEG, EMG);
- pressures of liquids (blood) and gases (respiratory gas);
- flows and volumes of liquids (the stroke volume and cardiac output of the heart) and gases (the tidal volume and minute volume of the lung);

[1] Ream AK. Monitoring concepts and techniques. In: Ream AK, Fogdall RP (eds). Acute cardiovascular management; anesthesia and intensive care. Lippincott, New York. 1982.

- compositions of liquids (blood) and gases (respiratory gas);
- frequencies or periods (heart rate, respiration frequency);
- temperatures (core temperature, skin temperature).

The text discusses techniques to measure most of these variables, why they should be measured and what they tell about the patient. It is not complete, in the sense that little will be said about existing measurement devices. Its goal is, foremost, to describe measurement *principles* and the anatomy and physiology on which the principle is based, because these are important for the measurement's *interpretation*. A number as such tells us nothing; it needs an interpretation, and the interpretation may lead to an action.

The text is written for science and engineering students. It assumes that the reader has little knowledge of biology, medicine, anatomy and physiology, and where needed it provides an introduction into some of the concepts of these sciences. It further assumes that the reader is well versed in algebra and at least the calculus of differential and difference equations. Moreover, many of the models are given in terms of electrical networks, with which electrical engineering students are most familiar. One reason for this is that the literature uses these models as well. Although the derivations also could be done in terms of mathematical (differential) equations or in terms of networks of springs and dampers with which mechanical engineers are more familiar, electrical networks provide more succinct models and remain closer to the existing literature. Another reason is that simulation software exists (e.g., SPICE) which, although designed for electrical networks, is also well suited for simulations of the biomedical models that are treated in this text.

This text should preferably be read more than once. The first two chapters provide a broad overview, whereas later chapters are more specific. It helps understanding to reintegrate those specifics into the broad picture on a second reading of the early chapters. Some of the concepts that are treated in early chapters are explained more fully in later chapters. Chapter 1, for instance, introduces "airway resistance" as an important notion that significantly aids the understanding of certain phenomena, whereas Chapter 2 shows that this concept is only an idealization that may be quite inappropriate in some cases. Also, although it may be possible to isolate one aspect of the patient in a certain measurement, that single aspect can usually only be understood when it is related to other aspects. That statement is valid for this text as well: each detail may need to be related to many other details. It is only when these relations are routinely perceived that understanding has been reached.

In many cases, the presented information is sketchy. A complete book could be (and probably has been) written about any of the presented subjects. For a deeper understanding, the reader must be referred to more detailed books or to a www-search which, when given the correct search terms, delivers a wealth of information: introductory texts, discussions about details, medical and technical papers, simulations, and currently existing devices. The goals of this text are a survey of the field, a review of the necessary fundamentals on which deeper study can be based and an overview of possible search terms. The courses that were based on this text were traditionally combined with a course in literature research; that is why this text does not contain references. Two books that have proven to be excellent primary references need to be mentioned, however:

A.C. Guyton and J.E. Hall: *Textbook of Medical Physiology*, 10th ed. London: Saunders, 2000. ISBN 0-7216-8677-X.

J.D. Bronzino: *The Biomedical Engineering Handbook*, 2nd ed. Boca Raton, FL: CRC Press, 2000. ISBN 0-8493-0461-X (vol. 1), 0-8493-0462-8 (vol. 2).

If numbers of physiological variables are given in the text, they invariably — unless otherwise indicated — refer to young healthy male adults at rest. These numbers may be very different in babies, the elderly or the sick, and during exercise.

We will frequently use *models* as an aid in understanding. It must be repeated that every model is a *simplification*; it focuses on what is deemed important and leaves out less vital details. Models are *context-dependent*, because the context usually determines what is important and what not. Choosing the correct model is often a matter of much experience, and careful matching between the properties of the model and the real system that it models (*model validation*) is always required. Different models of the same phenomenon in which different details are neglected will result in different outcomes. In Chapter 2, for instance, we derive the "resistance" of a liquid-filled tube using two different models, that of Poiseuille and that of Womersley. The "Womersley resistance" turns out to be 1.5 to 3 times larger than the "Poiseuille resistance." This probably indicates that neither model is very accurate. It also indicates, however, that the term "resistance" is a useful one, although in practice its value may need to be acquired from actual measurements rather than from knowledge contained in a model.

The figures are sketches, models, simplifications, as well. Their simplicity is meant as a constant reminder that it is always models, simplifications and approximations that are discussed. If an approximation is within about 5% of the real thing, the accuracy is usually sufficient for purposes of monitoring in patient management.

Chapter 1 gives the elementary facts about the anatomy and physiology of the respiratory and circulatory systems that will be referred to time and again in later chapters.

Chapter 2 reviews the physics of fluid transport in "tubes" which will be applied to airways and blood vessels. This chapter also discusses why simplifications are necessary for a good understanding of the basic physiological phenomena.

Chapters 3 to 7 present methods to acquire physiological measurement and the models that they are based on.

Chapter 8 presents some therapeutic devices that are often used in patient monitoring, which can thoroughly change the appearance of measured physiological signals.

Chapter 9 discusses how individual measurements are combined into an overall picture of the condition of the patient, how physicians use that overall picture as a basis for their therapeutic actions and how therapy in turn influences the measurements. This chapter thus completes the presentation of patient monitoring as a feedback process.

Acknowledgments

This text is partly based on earlier versions of materials that were used for many years in courses offered to engineering students by the Department of Electrical Engineering of the Eindhoven University of Technology in the Netherlands. Authors of these earlier versions were J.E.W. Beneken, W.H. Leliveld and M. Stapper. Suggestions as to changes and additions from K.H. Wesseling and several biomedical engineering researchers at the Department of Anesthesiology of the University of Florida in Gainesville, J.J. van der Aa, S. Lampotang, W.L. van Meurs and H. van Oostrom, are gratefully acknowledged, as are the contributions of the many students who critically read the text, identified mistakes and omissions, and suggested improvements.

Contents

1 Basic Anatomy and Physiology

In anatomy, the form of the organs of the body is studied; the basic question is "What does it look like?" In physiology, their function is studied; the basic question is "What does it do?" In modeling, form and function go together to a large degree, and since we want to focus on the major findings, we will freely mix anatomy, physiology and models (simplifications of anatomical and physiological conclusions).

Physiology views a human being as a complex system, where the major "goal" of the system "human" is the continuance of the intact individual. This "survival" can be viewed at different levels of complexity: the total system, its subsystems (organs), etc., and at the lowest levels the individual cells or even their components. The survival of the cells depends on a large number of factors, but the most time-critical ones are an adequate *supply of oxygen* and an adequate *excretion of carbon dioxide*.

Physiological functions can be grouped into a number of (sub)systems that can be more or less clearly distinguished. Some of these are the respiratory system, which brings oxygen into the body and removes carbon dioxide; the circulatory system, which transports the oxygen-enriched blood to the tissues and the oxygen-depleted blood to the lungs; the oxygen transport system, which allows the blood to carry more oxygen than would otherwise be possible; and the nervous system, which must take care of the integration of all subfunctions. The greatest and most acute threat to the human system is an inadequate oxygen supply to the brain, the body's most vulnerable organ. Three minutes without oxygen can lead to irreversible damage or death. Since we may assume that the brain will survive as long as respiration, circulation and oxygen transport are adequate, these physiological systems can be considered the most critical ones.

1.1. OVERVIEW OF THE RESPIRATORY AND CIRCULATORY SYSTEM

In surprisingly many ways, the needs of each cell in the human body resemble the needs of a unicellular organism floating in the ocean. Cells are highly organized membrane-enclosed factories, in which all sorts of chemical processes take place that serve the cell's survival, growth and reproduction. Living cells require a continuous supply of organic and inorganic materials (e.g., O_2, water, glucose, protein, Na^+, K^+, Ca^{++}) and discharge other materials to their surroundings (e.g., CO_2, lactic acid; also heat). But whereas a unicellular organism is so small that the uptake of the necessary chemicals from its environment and the elimination of its waste products back into the environment can largely be done by diffusion through its cell membrane, the cells of the human body are packed so densely together that diffusion as the sole transport mechanism is inadequate. Diffusion in organisms larger than a few hundred microns would be intolerably slow (in a human diffusion from head to toe would take about 100 years). Multicellular organisms therefore need additional mechanisms that bring individual cells into contact with the outside environment. The respiratory and circulatory systems together are such a mechanism; their main function is to transport oxygen — required to "burn" the cell's "fuel" (which ultimately comes from the food that we eat) — to the individual cells and to release carbon dioxide — the waste product of the "burning" process — back into the environment. Respiration and circulation together ensure that each individual cell is surrounded by extracellular (or "interstitial") fluid of just the right composition.

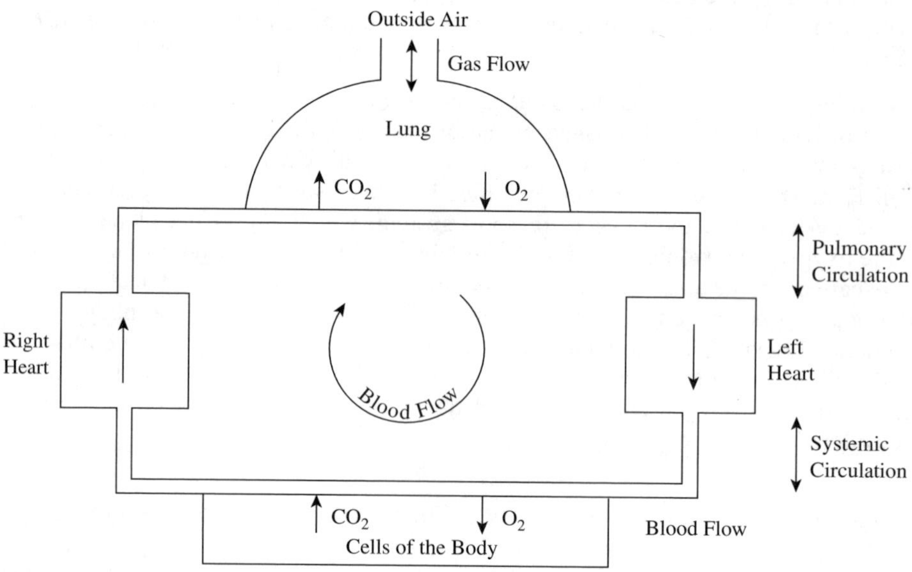

Figure 1.1. The task of the respiratory and circulatory systems is to transport oxygen from the outside air to the peripheral cells and to transport carbon dioxide back from the cells to the outside air.

The diagram of Figure 1.1 gives a summary of how this is done. Air,[1] which contains about 21% oxygen, enters the lung.[2] The oxygen in the lung diffuses to the surrounding blood vessels. The heart consists of two pumps, combined in one package. The right part of the heart pumps the blood through the vessels that surround the lung so that oxygen can diffuse into the blood (the blood is "oxygenated"); this part of the circulatory system is called the *pulmonary* circulation. Next, the left part of the heart pumps the oxygen-enriched ("red") blood through the

[1] With *air*, gas with the normal atmospheric gas composition is meant. The term "respiratory *gas*" will also be used because patients often receive gas with a different composition.
[2] Although anatomically we have two lungs, we will speak of "the lung" as the organ that takes care of respiration.

tissues of the body; this part of the circulation is called the *systemic* circulation. Here oxygen diffuses to the cells and the carbon dioxide produced by the cells diffuses back into the blood. The oxygen-depleted and carbon dioxide-enriched ("blue") blood is now transported back to the lung where the carbon dioxide diffuses from the blood to the lung, from where the carbon dioxide is transported to the environment. This process continues as long as we live; indeed, when this process somehow stops, we die.

We can subdivide this whole process into four subprocesses, which we will consider in turn.

Respiration and ventilation are mechanical processes that transport gas back and forth between lung and environment. Respiration (spontaneous breathing) requires muscular effort. When a patient's muscles are artificially put out of order ("relaxed" by a drug, e.g., for surgery), the patient needs to be *ventilated* (artificially respirated). To understand respiration and ventilation, we need to study the mechanics of gas flows, pressure differences and volume changes. This analysis is called *lung mechanics*.

Diffusion is the passive physical process that exchanges oxygen and carbon dioxide between the alveoli and the blood (in the lung), and between the blood and the cells (in the periphery). All gas exchange is by diffusion. To understand diffusion, we need to study partial gas pressures in alveolar gas and in blood, and the lung's diffusion capacity.

Circulation is the mechanical process that transports blood from the lung to the tissues and back through the pump action of the heart, so that oxygen can be transported to the cells and carbon dioxide back to the environment. In order to understand the circulation, we need to know how the heart pumps blood and how the blood circulates due to pressure differences.

Gas transport in the blood is a process that transports oxygen and carbon dioxide from the lung to the individual cells and back. This subprocess depends on the properties of the blood, and particularly on the chemical reactions that take place when oxygen and carbon dioxide are buffered (see Section 1.3.2).

1.2. ANATOMY AND PHYSIOLOGY OF THE LUNG AND THE AIRWAYS

Figure 1.2 shows a horizontal cross section of the thorax (chest) at the height of the lung. The two lungs, which take up most of the cross-sectional area, are enclosed by an elastic coating called the visceral pleura, while the inside of the ribs, which together form the thorax, is covered by a similar coating called the parietal pleura. The *intrapleural space*, the space between both pleurae,[1] is filled with a thin layer of fluid which serves as a lubricant that allows frictionless movement of the pleurae

[1] The space between both pleurae is called the *intrapleural* space (or *pleural* space) because both pleurae are actually one and the same anatomical structure, one pleura folding back upon the other.

with respect to each other when the volumes of lung and thorax change during respiration. If this fluid is missing or has an inappropriate composition, respiration becomes very painful.

The space between both lungs is occupied by the heart, the two main bronchi that transport gas to and from both lungs, and the esophagus which leads to the stomach.

When we analyze the anatomy of airways and lungs, carefully detached out of a body, the small size of the lungs is surprising. They have shrunk considerably from the size they had when still in the thorax (about 6 l). But the small size does not prevent an anatomical analysis.

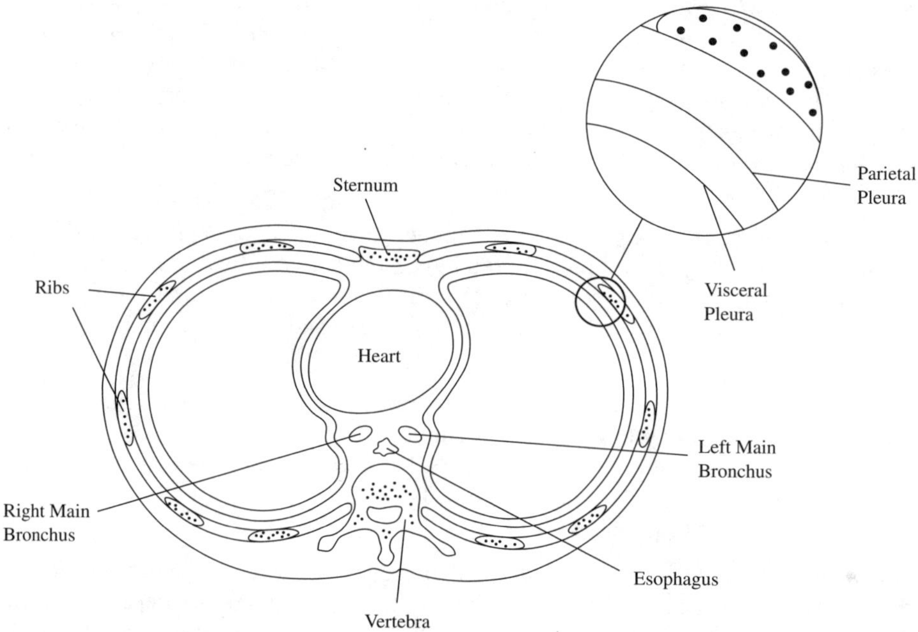

Figure 1.2. Horizontal cross section of the thorax at the height of the lung. Ribs run from the spinal column in the back to the sternum at the front of the thorax. The two main bronchi, the esophagus and the heart occupy the space between both lungs. The space between the visceral pleura and the parietal pleura is called the *intrapleural* space.

1.2.1. Trachea, bronchi, alveoli

Gas enters and leaves the lung through a bifurcating system of tubes that get successively smaller in diameter (Figure 1.3). Gas enters through mouth and/or nose and travels through the trachea, which branches into two bronchi ("airways"), one for each lung. These are called the main bronchi or the bronchi of the first generation (i.e., the bronchi after the first split). The bronchi subsequently branch time and again into ever smaller bronchi, up to the 11th generation. From then on, the tubes are called bronchioles ("small bronchi"), up to the 19th generation. Ultimately, in the 23rd generation, the alveoli are reached, small air bubbles that form the "leaves" of the "respiratory tree." The alveoli and the small tubes of the highest several generations (called respiratory bronchioles and alveolar ductuli) comprise most of the lung's volume. Since the number of tubes doubles in each generation, there are about 2000 bronchi of generation 11 and about 8,000,000 alveoli. This tremendous branching is required in order to provide a sufficiently large surface (about 50 to 100 m^2) through which diffusion can take place.

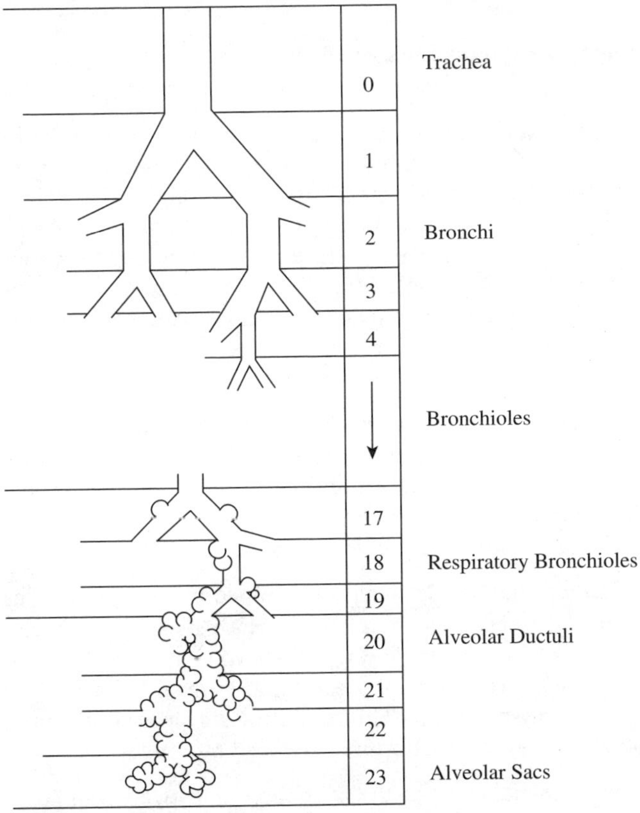

Figure 1.3. The generations of the bronchial tree. Generations 1 to 16 transport gas. Gas exchange takes place in generations 17 to 23.

Question: Assume a lung volume of 6 l. If we model both lungs together as one perfect sphere, what would the lung surface area be? And what would the lung surface be if this one perfect 6-l sphere were densely packed with 8,000,000 small perfect spheres?

Generations 1 through 16 are called the *conductive zone*; they mainly serve as airways that transport gas to and from generations 17 through 23, the *respiratory zone*, where oxygen and carbon dioxide exchange takes place between the air in the lung and the blood in the small-diameter blood vessels that surround the alveoli. The air ducts of generation 0 (the trachea) through generation 11 are surrounded by cartilaginous rings, which prevent their collapse. Higher generation ducts are so compliant that they collapse (are compressed) when the pressure outside a duct becomes larger than the inside pressure, i.e., when the *transmural pressure* becomes negative (see Section 2.9). The transmural pressure is defined as the pressure *inside* the tube minus the pressure *outside* the tube. In an infinitely compliant tube, the transmural pressure cannot become negative; the outside pressure would just close the tube. Normally, the transmural pressure can become slightly negative before the tube is fully closed.

1.2.2. Surfactant, intrapleural pressure

The lung tissue is elastic, and the lung by itself would — like a balloon — tend to decrease its volume. Since the fluid in the space between both pleurae cannot expand, this elastic force is in turn propagated to the thorax. The thorax, however, is also elastic and also tends to preserve its shape and its volume. As a result of both pulling forces, a negative *intrapleural pressure* of approximately 5 cm H_2O[1] at rest arises in the intrapleural space. If, somehow, a perforation causes the intrapleural space to become connected to the outside air, it loses its underpressure and both lung and thorax will assume their resting volume. This condition is known as pneumothorax.

We will now be more specific and study the origin of the intrapleural pressure in detail. Let us start with a model for a single alveolus. The alveolus contains air and we can think of it as being surrounded by liquid, since its surface consists of a thin layer of cells with an embedded network of tiny capillaries. This prompts the analogy of an alveolus with an air bubble in water.[2]

The volume of an air bubble in water is determined by two forces. The first force is due to cohesion and is characterized by a surface tension γ, which occurs at the separation between air and water. The surface tension acts to compress the air inside the bubble and thus produces an overpressure in the bubble relative to the hydrostatic pressure in the fluid. This overpressure causes an opposing force. A stable air bubble requires an equilibrium between both forces.

As a thought experiment, we split the air bubble into two equal parts (Figure 1.4). The surface tension γ tends to keep the rims of the upper and lower half bubbles together and thus acts around the *circumference* of the split bubble; the force due to it is

$$\gamma \, 2 \, \pi \, r \tag{1.1}$$

[1] The units mm Hg and cm H_2O are still very much in use in the medical community, the first especially for circulatory pressures and the second for respiratory pressures. We will use these units too. It helps to remember that 1 kPa = 10 cm H_2O = 7.5 mm Hg.

[2] This is basically how models come into existence: a new situation is expressed in terms of one already known. Whether that model is accurate enough to represent the new situation is still to be tested, a process called model validation.

where r is the air bubble's radius. The overpressure ΔP tends to drive the two half bubbles apart and thus acts over the *surface* of the cut; the force due to it is

$$\Delta P \, \pi \, r^2 \quad (1.2)$$

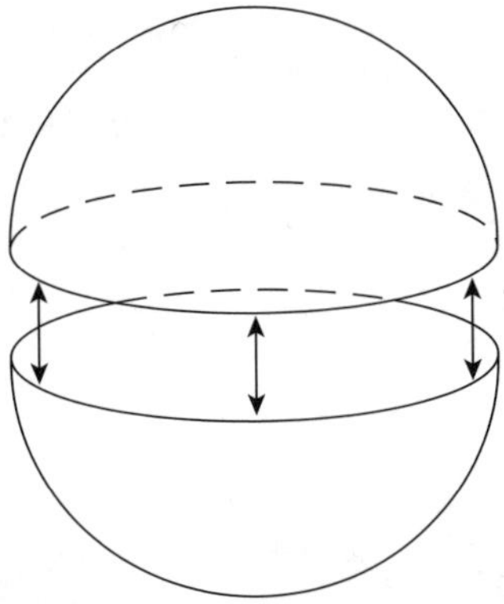

Figure 1.4. Two forces determine the volume of a gas bubble under water. Surface tension acts at the gas-liquid interface and attempts to decrease the volume. The result is an over-pressure inside the bubble, which resists the volume decrease.

For a stable bubble we have an equilibrium between the two

$$\gamma \, 2 \, \pi \, r = \Delta P \, \pi \, r^2 \quad (1.3)$$

or

$$\Delta P = 2 \, \gamma \, / \, r \quad (1.4)$$

The latter expression is known as Laplace's law. It indicates that high pressures will exist in small bubbles. The lung, however, consists of alveoli of unequal sizes. Laplace's law now tells us that the pressures in different sized alveoli will be unequal, the smaller alveolus having the greater pressure. Since alveoli are mutually interconnected via airways, the pressure difference would cause airflow. This flow would go from the higher pressure (the smaller bubble) to the lower pressure (the larger bubble), causing the smaller bubble to empty itself into the larger one and thus cease to exist. This is clearly not the case in a healthy lung.

That even the smallest alveoli of a healthy lung can remain in existence is due to a liquid called *surfactant*, a product of the alveolar cells, which covers the inside of the alveoli in a mono-molecular layer. Surfactant, a soap-like surface-active material, causes the surface tension to become directly dependent on the alveolus' radius, and it also provides for a hysteresis that tends to keep the alveolus' radius constant (Figure 1.5). The effect is that alveoli of different radii can coexist peacefully.

For an alveolus with a radius of 0.2 mm, the surface tension would be about 40.10^{-3} N/m. Thus

$$\Delta P = 2 \, \gamma \, / \, r = 2 * 40.10^{-3} \, / \, 0.2.10^{-3} = 0.4 \text{ kPa} = 4 \text{ cm } H_2O \quad (1.5)$$

which means that the transmural pressure (across the wall of the alveolus) must have a value of 4 cm H_2O, with the higher pressure inside the alveolus. Equivalently, a pressure of –4 cm H_2O on the outside would keep an alveolus whose internal pressure is zero (because the airways provide an open connection to the outside world) open. Macroscopically, this is the situation with the whole lung which is kept open by the underpressure in the intrapleural space. The intrapleural pressure depends upon lung volume and can range from about –2 to about –15 cm H_2O.

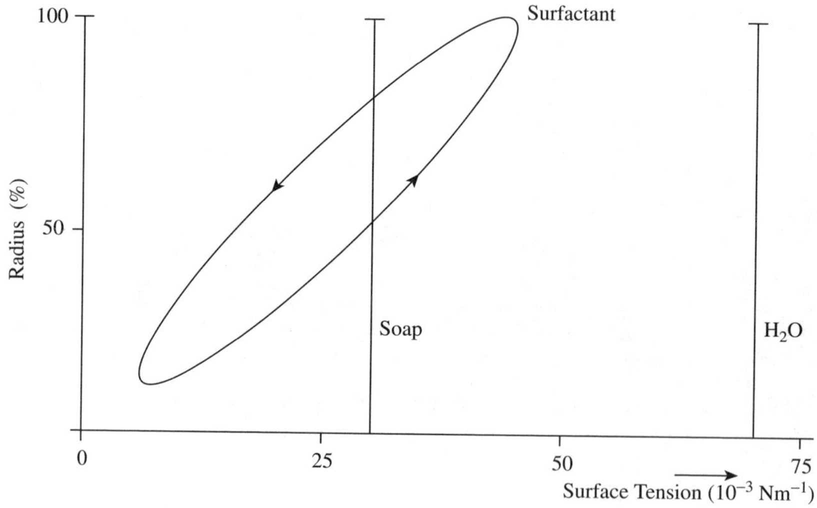

Figure 1.5. Because of the surfactant, there is an approximately linear relation between the radius of an alveolus and its surface tension. For reference, the surface tension values for soap and water are given as well.

1.2.3. Volume-pressure relationships of lung and thorax

We have already seen that the lung consists of elastic tissue and that a pressure is required to keep it open, the normal transmural pressure being about –5 cm H_2O. In the context of lung mechanics, a balloon is often taken as a model for the lung. Just like with a balloon, we can learn more about the volume-pressure relationship of a lung if we slowly increase the volume of an isolated lung and keep track of the resulting pressure. We can also determine the volume-pressure relationship of an isolated thorax. Figure 1.6 shows the results, as well as the volume-pressure relationship for the lung inside — in combination with — the thorax, which also simply results from adding the two individual curves. In fact, the combined curve can be measured in a cooperative patient, who must not breathe during the test and keep the glottis open, by slowly adding volume to the lung.

Point A in Figure 1.6 indicates the resting volume of the thorax, i.e., the volume the thorax would have if the pressure across the thorax wall were zero and if there were no muscle action. Point B is the resting volume of the combination of lung and thorax, which exists at the end of a normal expiration. This volume is called the functional residual capacity, FRC (see Section 1.2.5).

Question: Figure 1.6 assumes that we apply a volume and measure the resulting pressure. We could also apply a pressure and measure the resulting volume. Redraw the figure with pressure as the independent variable.

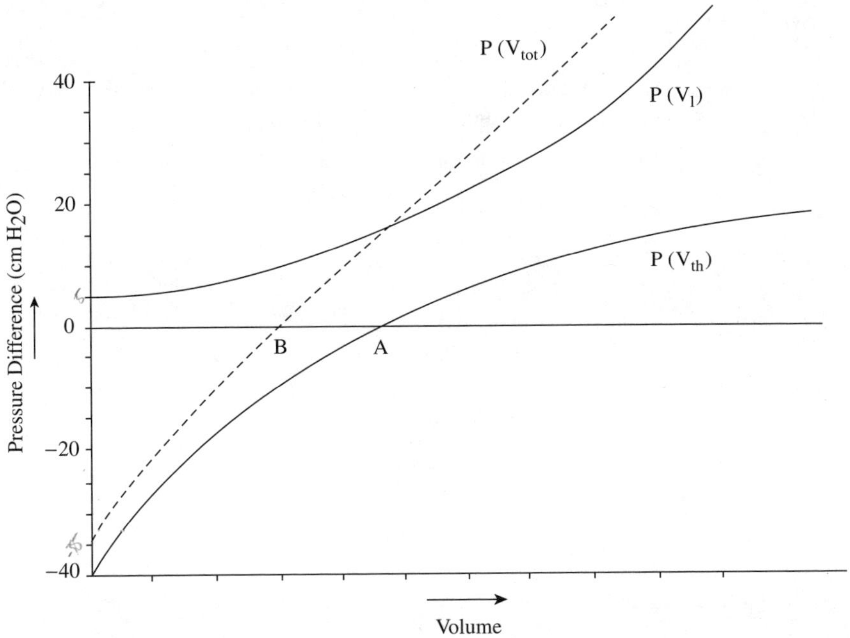

Figure 1.6. Pressure-volume relationships of lung V_l, thorax V_{th} and the combination of both V_{tot}. The volume scale has markers at 1-l intervals. Volume A is the resting volume of the thorax. The smaller volume B is the resting volume of the lung-thorax combination.

Figure 1.6 also shows that, when volumes and pressures do not change too much — as in normal breathing or normal ventilation — all curves are reasonably linear around their normal working point. This is true for all lungs. The slopes of the lines indicate *compliances*, where a compliance is defined as $C = \Delta V / \Delta P$, that is, a volume change resulting from a pressure change. Lung compliance is determined by the mechanical properties of the lung's elastic connective tissues and by the surface tension of the fluid on the alveolar walls. The latter is greatly reduced (and thus the compliance increased) by surfactant. A small compliance indicates a stiff lung, a large compliance a compliant lung. Normal values of the compliances of the lung and of the thorax are both around 0.13 l/cm H_2O. The *lung-thorax compliance* is the compliance of lung and thorax combined.

Breathing causes volume changes of the thorax. An inspiration is caused by muscle action which creates an addition to the thorax volume and can thus be represented by shifting the curve of the thorax volume to the right, whereas a forced expiration shifts it to the left. As a result, the combination curve also shifts. A linearized representation is given in Figure 1.7. It shows that on inspiration the resting curve shifts downward. This shift can be interpreted as either a volume change where the pressure change is zero (the chest expansion) or a pressure change at the same volume (due to muscle action).

We can write these relationships as

$$V_{thorax} = C_{thorax}\,(P_{thorax} + P_{muscle}) = C_{thorax}\,(P_{ip} + P_{muscle}) \tag{1.6}$$

$$V_{lung} = C_{lung}\,P_{lung} = C_{lung}\,(P_{alv} - P_{ip}) \tag{1.7}$$

where P_{muscle} is the pressure caused by the action of the respiratory muscles, P_{ip} is the intrapleural pressure and P_{alv} is the pressure in the alveoli.

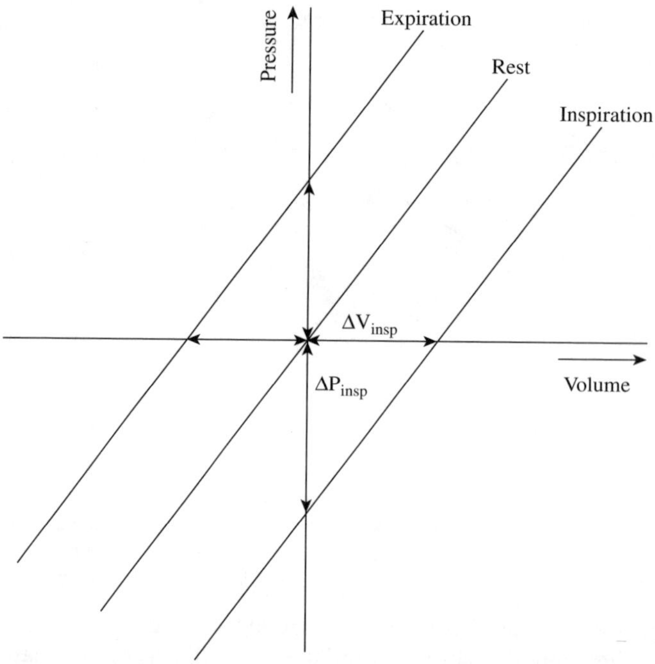

Figure 1.7. Linearized pressure-volume relationship of lung-thorax combination.

We now assume that $V_{thorax} = V_{lung} = V$ (which is true because the intrapleural space has negligible volume) and that $C_{thorax} = C_{lung} = C$ (which is usually approximately true). Then the above formulas tell us that

$$P_{ip} + P_{muscle} = P_{alv} - P_{ip} \tag{1.8}$$

or

$$P_{ip} = \tfrac{1}{2}\,(P_{alv} - P_{muscle}) \tag{1.9}$$

The pressure difference between alveoli and outside air is thus

$$P_{alv} = V\,/\,(\tfrac{1}{2}C) - P_{muscle} \tag{1.10}$$

The term ½C stands for the compliance of lung and thorax together, also called the *lung-thorax compliance* $C_{lung\text{-}thorax}$. The lung volume is thus

$$V = C_{lung\text{-}thorax}\,(P_{alv} + P_{muscle}) \tag{1.11}$$

Question: Derive the expression for the lung volume when both compliances are not equal. Note that this question can be rephrased as: How can $C_{lung\text{-}thorax}$ be computed if C_{lung} and C_{thorax} are given.

There are thus two ways to change the volume of the lung, through P_{alv} and through P_{muscle}. When the airways are open and breathing is slow, the pressure drop due to the airways resistance can be neglected and the pressure P_{alv} in the lung will be approximately zero. A change of P_{muscle} will directly cause a volume change

$$V = C_{lung\text{-}thorax}\,P_{muscle} \tag{1.12}$$

When a patient is ventilated, the muscles are usually relaxed and $P_{muscle} = 0$. Volume changes can then be induced by presenting a positive pressure P_{alv} to the lungs

$$V = C_{lung\text{-}thorax}\,P_{alv} \tag{1.13}$$

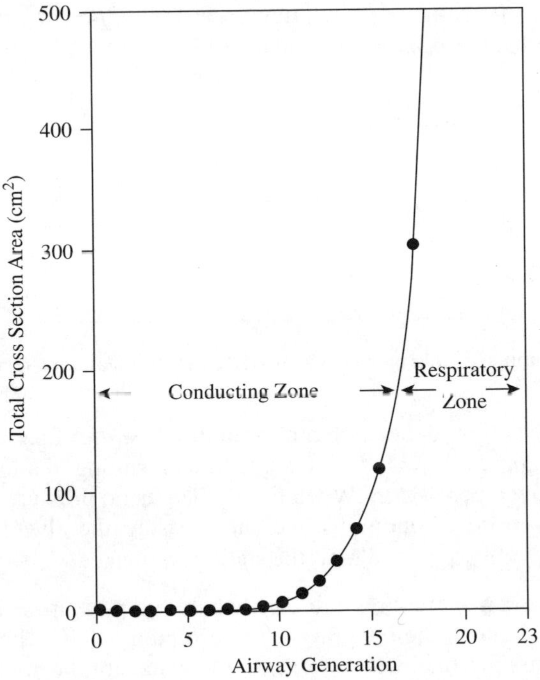

Figure 1.8. Total cross-sectional area of each airway generation. The resistance to flow is mainly concentrated in generations 0 to about 12.

This, too, assumes that the breathing is slow and that airway resistance can be neglected.

A pressure difference ΔP along a section of the airways causes a flow F through that section. The friction between gas and airway walls causes an opposition to the flow, which is called the airway *resistance* R (see Section 2.3):

$$R = \Delta P / F \tag{1.14}$$

Note the resemblance of this equation to Ohm's law.

Airway resistance depends upon the type of flow (laminar or turbulent) and on the cross sections of the tubes through which the gas flows. Numerous small airways in parallel provide less resistance to a certain flow F than one large airway with an equivalent cross section (see Section 4.1.1). As a result of the tremendous branching of the airways and the rapid increase of the total cross-sectional area at higher generations (Figure 1.8), the airway resistance is mainly due to the upper airways, trachea and bronchi. Because the airways distend as the lung becomes larger, the airway resistance decreases as lung volume increases.

The simplest lung mechanics model consists of only airway resistance R and lung-thorax compliance C (Figure 1.9). It assumes that the patient's respiratory muscles are inactive and that external equipment (a ventilator, see Section 8.2) is used to insufflate the lungs by imposing a flow F. Since the alveolar pressure P_{alv} is not accessible, we will use the pressure P_M measured at the patient's mouth instead. Due to the pressure drop across R, we will now measure a *different* compliance $C = \Delta V / \Delta P_M$, which is called the *dynamic* compliance.

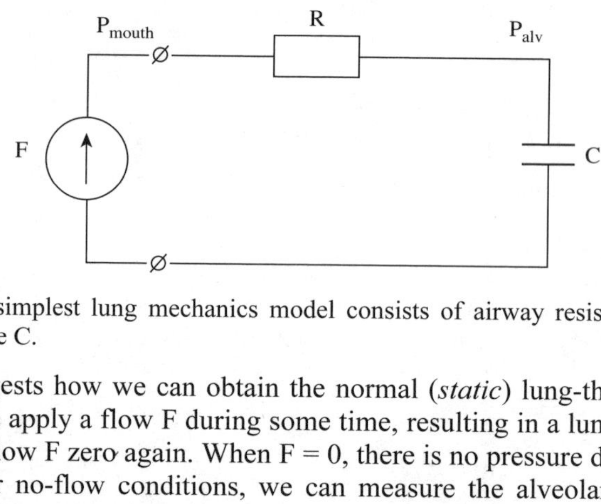

Figure 1.9. The simplest lung mechanics model consists of airway resistance R and lung-thorax compliance C.

Figure 1.9 suggests how we can obtain the normal (*static*) lung-thorax compliance of a patient. We apply a flow F during some time, resulting in a lung volume V, and then make the flow F zero again. When F = 0, there is no pressure drop across R and $P_M = P_A$; under no-flow conditions, we can measure the alveolar pressure at the mouth and $C = \Delta V / \Delta P_M = \Delta V / \Delta P_A$, the static compliance.

Figure 1.10 shows a volume-pressure plot, where the pressure P is measured at the mouth and the volume is the integral of the respiratory flow. The dynamic compliance C_{dyn} is the straight line, which forms the long axis of the approximately elliptic curve.

The area enclosed by the curve of Figure 1.10 is called the *(physiologic) work of breathing*; it is the resistive work performed by the ventilator (and when the patient breathes spontaneously, by the patient) to overcome the resistance that is present in the airways. There may also be *imposed* work of breathing if resistance R is artificially increased, e.g., by the resistance inherent in the tubes and hoses of the ventilator and the endotracheal cuff. Imposed work of breathing should be minimal, especially in spontaneously breathing patients with a compromised lung function, since it causes increased respiratory muscle loading and oxygen consumption, which may result in respiratory muscle fatigue and possibly even failure.

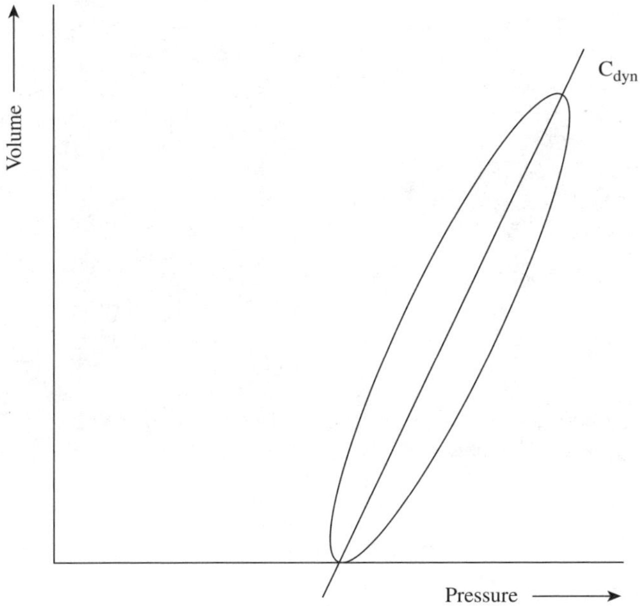

Figure 1.10. A respiratory PV-loop. The dynamic compliance C_{dyn} is the major axis of the ellipse.

1.2.4. The respiratory muscles

Spontaneous respiration is an active process, in which contractions of respiratory muscles cause a volume change of the thorax and thus a volume change of the lung. When breathing normally, only inspiration is active. The ribs are elastically connected (through cartilaginous joints) to the vertebrae of the spine and to the sternum at the front of the chest.

When a group of *intercostal muscles* (the muscles between the ribs) contract, they force the ribs apart and thus lift the rib cage as a total, increasing its volume; this is called chest breathing. The diaphragm, a large muscular plate- or dome-like structure that separates thorax and abdomen, contracts as well, which also increases the volume of the thoracic space; this is called abdominal breathing. Both volume changes are about 250 ml in an adult, combining into a *tidal volume* (the volume of one breath) of about 500 ml.

Exhalation is normally a passive process; the respiratory muscles relax and, due to the elasticity (compliance) of both lung and thorax (and gravity forces, when erect), lung and thorax reassume their resting volume. Exhalation can be speeded by another group of intercostal muscles, whose effect is to forcefully reduce the thorax volume. Figure 1.11 shows the positions of the respiratory muscles, and Figure 1.12

schematically shows the directions of the forces resulting from the contractions of the intercostals.

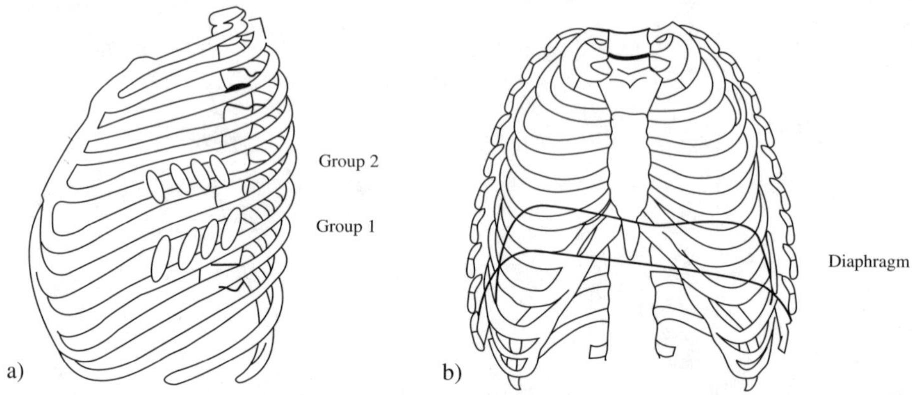

Figure 1.11. The respiratory muscles: a) inspiratory intercostals (group 1); expiratory intercostals (group 2); and b) diaphragm.

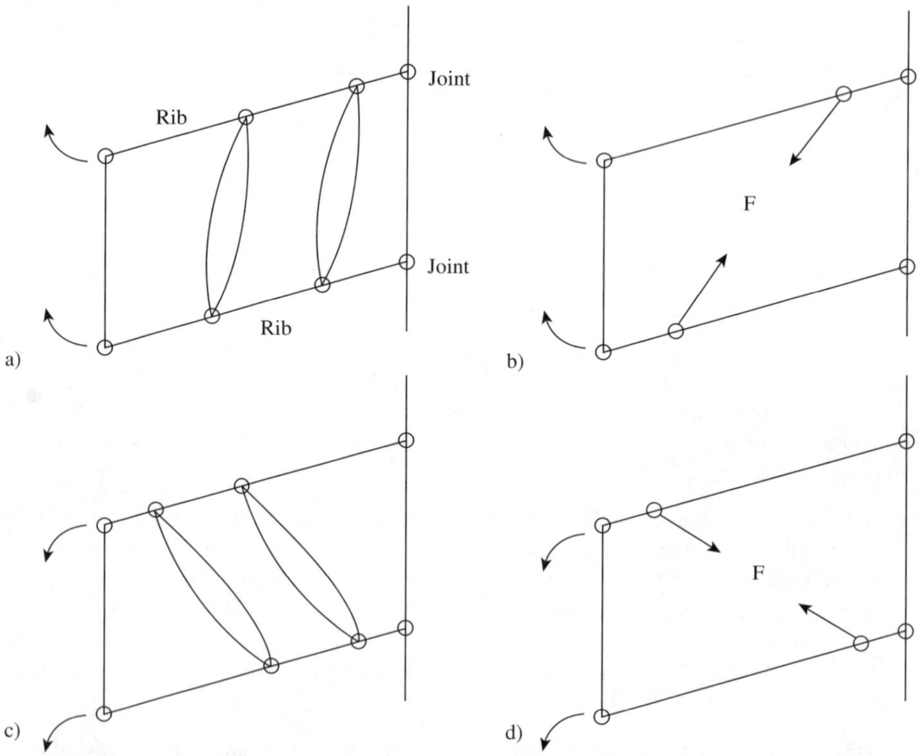

Figure 1.12. Inspiration: the inspiratory intercostal muscles force the ribs apart and lift the thorax forward and up (a). Expiration: the expiratory intercostals force the ribs together and move the thorax backward and down (c). The rigid segments models (b for inspiration; d for expiration) show the directions of the forces of both muscle groups.

1.2.5. Lung volumes

We have already encountered some names for different lung volumes. Pulmonary medicine employs the following terminology for different volumes (Figure 1.13; values are for a healthy adult):

- The *Total Lung Capacity* (TLC) is the maximum volume of gas that can be contained in the lung through voluntary effort (about 6 l). This volume is realized at the end of a maximum inspiration.
- The *Vital Capacity* (VC) is the maximum volume of gas that can be exhaled after a maximum inspiration (75 to 80% of TLC).
- A certain volume of gas cannot be exhaled and will always remain in the lung; this volume is called the *Residual Volume* (RV, 20 to 25% of TLC).
- The *Inspiratory Capacity* (IC) is the maximum volume of gas that can be inhaled from the resting level after a normal expiration (about 60% of TLC).
- The *Functional Residual Capacity* (FRC) is the gas volume that remains in the lungs after a normal expiration (about 40% of TLC).
- The *Inspiratory Reserve Volume* (IRV) is the maximum gas volume that can still be inhaled after a normal *in*spiration (about 50% of TLC).
- The *Tidal Volume* (TV) is the gas volume that is inhaled and exhaled during normal, quiet breathing (0.5 l, about 10% of TLC).
- The *Expiratory Reserve Volume* (ERV) is the maximum gas volume that can still be exhaled after a normal *ex*piration (15 to 20% of TLC).

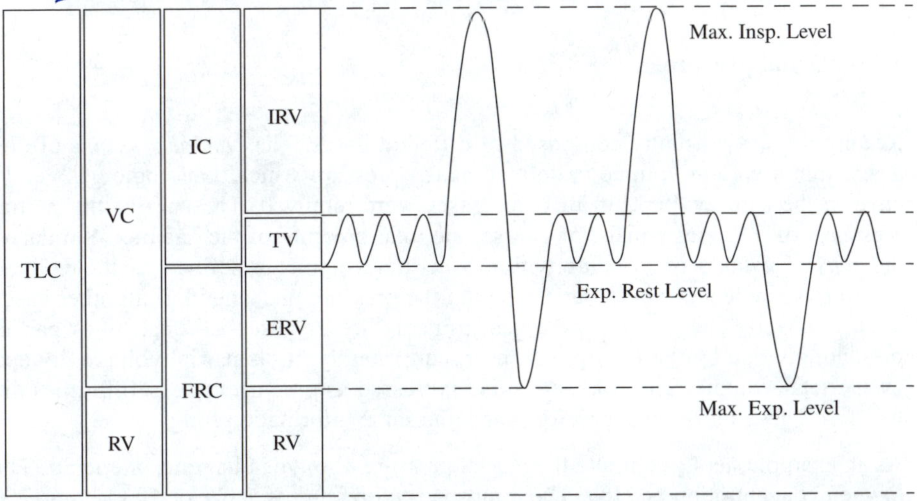

Figure 1.13. At the left, the different lung volumes are indicated. The right part shows a time trace of lung volume as the patient breathes quietly, inhales fully up to the maximum inspiratory level, then exhales fully down to the maximum expiratory level.

Other terms that are in use:

- The *respiration frequency* is the number of breaths per minute (normally 10 to 12).
- The *minute volume* is the product of tidal volume and respiration frequency (about 6 l/min).
- The *alveolar ventilation* is that part of the minute volume that reaches the alveoli so that it can participate in gas exchange (about 4.2 l/min). The remaining part of the minute volume (about 1.8 l/min) does not reach the alveoli but remains in the large airways that form the conducting zone, where no gas exchange takes place. This zone is called the *anatomical dead space* (normally about 150 ml).
- The *alveolar perfusion* is the blood volume that perfuses the lungs each minute (about 5 l/min). In a healthy subject the right ventricle of the heart pumps all the blood through the lung; the alveolar perfusion is then equal to the *cardiac output*. If some of the blood takes a path that bypasses the lung, we have *shunt flow*; due to the shunt, some of the blood cannot acquire oxygen and release carbon dioxide.
- The *oxygen consumption* is the oxygen volume that is inhaled but not exhaled (about 250 ml/min). The *carbon dioxide production* is the carbon dioxide volume exhaled per minute (from 180 to 250 ml/min, depending upon diet). The *respiratory quotient* is the ratio of carbon dioxide produced and oxygen consumed; a normal value is 0.8.
- The *maximum ventilation* is the maximum gas volume that can be voluntarily inhaled and exhaled per minute (about 170 l/min).

1.2.6. Partial pressures

Respiratory air is usually composed of different gases. The *partial pressure* of one gas within a certain volume is defined as the pressure which that single gas would have in the same volume if all other gases were removed. The sum of the partial pressures of all gases equals, of course, the total pressure of the gas mix. Similarly, the *partial volume* of one gas within a certain volume is defined as the volume which this single gas would have, keeping the pressure unchanged, if all other gases were removed. The *fraction* of a gas in a mixture of gases is equal to its partial pressure divided by the total pressure or, equivalently, to its partial volume divided by the total volume. The *concentration*, expressed as a percentage, is 100 times the fraction. These four terms therefore carry the same information for gases.

As an example, let us compute the partial pressure of oxygen in atmospheric air. The oxygen concentration is 21%. The atmospheric pressure is 1 bar or 100 kPa or 760 mm Hg (1 mm Hg = 1 torr) or, to a good approximation, 1000 cm H_2O. The oxygen partial pressure is thus 21% of these numbers, that is, 160 mm Hg or 210 cm H_2O.

Boyle's law relates gas pressure and volume to temperature. It states that

$$P\,V = N\,R\,T \tag{1.15}$$

where P is pressure, V is volume, N is the number of moles of the gas under consideration (1 mole is the molecular mass of the gas), R is the universal gas constant and T is absolute temperature. An "ideal gas" is defined as a gas for which Boyle's law is valid. Some gases are more "ideal" than other gases. In practice, Boyle's law applies accurately only to gases with a boiling point far below room temperature. This includes many of the gases that play a role in respiration, such as oxygen, nitrogen, and carbon dioxide. It does *not* include vapors (fluids with a boiling point near room temperature), such as anesthetic "gases" like halothane or water vapor, which is also far from "ideal."

Boyle's law can also be written as

$$P = \rho\,R \tag{1.16}$$

where ρ is the gas *density* N/V.

Table 1.1. Water vapor partial pressures at various temperatures.

°C	mm Hg	cm H_2O	kPa
18	15	20	2.0
26	25	33	3.3
37	47	63	6.3

Boyle's law also applies to partial pressures and partial volumes of ideal gases; both are proportional to absolute temperature. But since water vapor is not an ideal gas, we need to take special measures, especially when we consider expiratory gas, which is water vapor saturated. One way is to physically remove all water vapor; this will leave ideal gases only (anesthetic vapors, if present, have concentrations of at most a few percent). Another way is to numerically correct for the partial pressure of water vapor. That compensation is necessary is shown in Table 1.1, which gives the partial pressures of saturated water vapor at different temperatures in three different units. In an ideal gas, a temperature increase from 18°C (291 K) to 37°C (310 K) would cause an increase of partial pressure by 310/291, that is, 6.5% higher, whereas for water vapor the same temperature increase causes a partial pressure increase of more than 200%.

What is even more important is what happens when a water vapor saturated gas mixture cools off. Since at lower temperatures the saturated water vapor partial pressure is less, some of the water vapor must turn into water. This explains why condensation will always take place in the expiratory tubing of ventilators. The condensed water will collect at the lowest point in the tubing and may, if excessive, cause expiration to become more difficult (see PEEP, Section 8.2).

In respiratory gas composition measurements, one always needs to specify volume, pressure, temperature and water vapor content. In order to compare measurements, certain standard conditions are introduced. The following terms are often encountered:

STPS: **S**tandard **T**emperature (0°C) and **P**ressure (1 bar), **S**aturated

STPD: **S**tandard **T**emperature (0°C) and **P**ressure (1 bar), **D**ry

BTPS: **B**ody **T**emperature (37°C), standard **P**ressure (1 bar), **S**aturated (63 cm H_2O)

ATPS: **A**mbient **T**emperature (18°C), standard **P**ressure (1 bar), **S**aturated (20 cm H_2O)

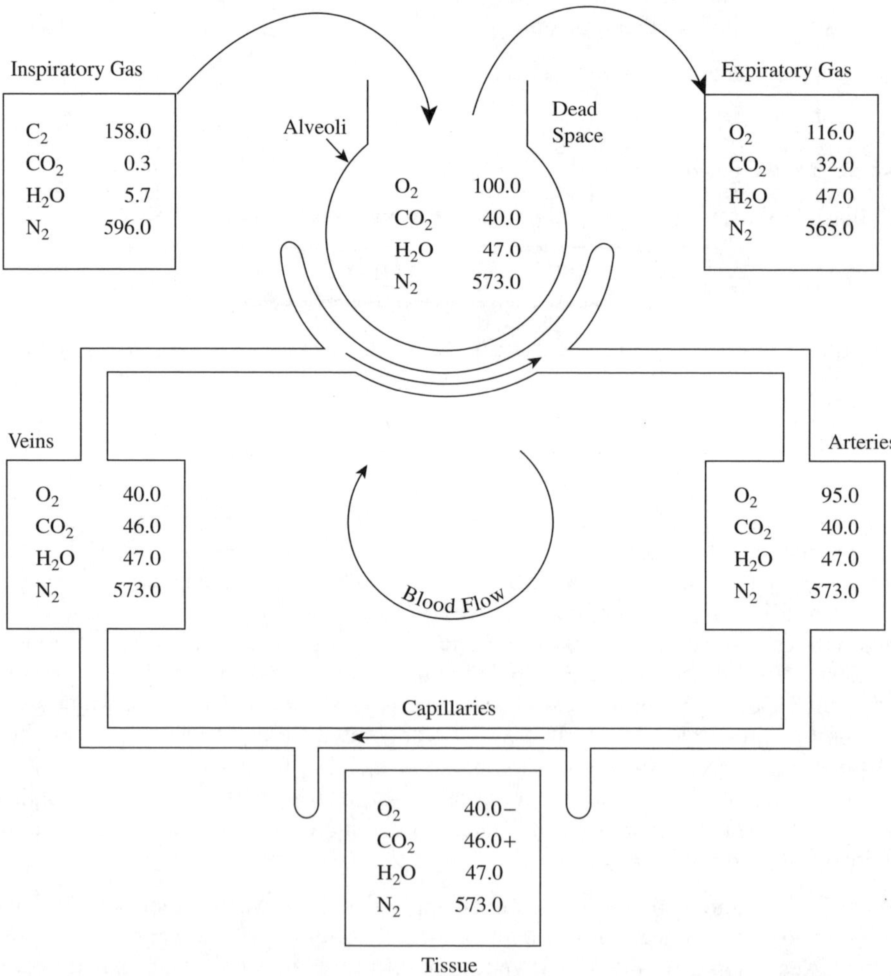

Figure 1.14. Partial gas pressures in mm Hg throughout the respiratory and circulatory systems. In some tissues, such as hard-working muscles, the oxygen partial pressure can go down to zero, with a concomitant increase in carbon dioxide partial pressure.

Question: Someone has stated that the oxygen partial pressure in expired gas is 100 cm H_2O BTPS. We want to know the equivalent partial pressure in cm H_2O STPD. What is it?

One also speaks of partial pressures of *blood gases*. When a gas is brought into contact with a liquid (e.g., water, blood), some of it will go into solution. When equilibrium has been reached — which may take some time, depending upon the solubility of the gas — the partial pressure of the gas dissolved in the liquid is by definition equal to the partial pressure of the remaining gas. The symbols for the partial pressures of oxygen and carbon dioxide in the alveoli are P_{AO2} and P_{ACO2}, the arterial ones are written P_{aO2} and P_{aCO2} and the venous ones are written P_{vO2} and P_{vCO2}.

Since alveolar gas and blood are so closely in contact, the partial pressures of oxygen and carbon dioxide in the blood leaving the pulmonary circulation will only differ slightly from the partial pressures in the alveolar gas. Figure 1.14 shows this: $P_{AO2} = 100$ mm Hg and $P_{aO2} = 95$ mm Hg, and $P_{ACO2} \approx P_{aCO2} = 40$ mm Hg.

Figure 1.14 also shows some other partial pressures throughout the respiratory and circulatory system. Let us consider the values for oxygen. Inspiratory gas (atmospheric air) contains about 21% O_2. Since atmospheric pressure is 760 mm Hg, the O_2 partial pressure is 21% of 760 mm Hg = 158 mm Hg. The alveolar gas concentrations, which can be measured at the end of expiration, show that some oxygen has disappeared. The O_2 partial pressure in the expiratory gas is higher than that of the alveolar gas due to mixing with the dead space O_2. Gas exchange has had almost enough time for an O_2 equilibrium across the alveolar membrane to be established. The arterial blood O_2 partial pressure is therefore almost as high as the alveolar O_2 partial pressure. The tissues extract oxygen, which is reflected in the O_2 partial pressures of the tissues and of the venous blood.

1.2.7. Dead space

The conductive zone of the airways, generations 1 through about 16, has a transport function only; it lacks capillaries and does not participate in gas exchange. Its volume is called the *anatomical dead space*. At the end of inspiration, this space has been filled with gas, but this gas is exhaled unchanged; its O_2 and CO_2 concentrations will be the same as those in the inspired gas. The anatomical dead space volume is about 150 ml, which means that about 30% of the air of a normal tidal volume of 500 ml is “unused”; although it takes effort (energy) to transport the dead space volume back and forth, none of its oxygen will be extracted.

In a diseased lung it can happen that certain parts of the lung are not perfused by the pulmonary circulation, although they are ventilated. These non-perfused parts of the lung are also dead space. The total dead space — anatomic plus non-perfused lung volumes — is called the *physiological dead space*. A significant dead space leads to respiratory problems, because wasted effort is expended to inhale and exhale the dead space gas volume. In a completely healthy person the physiological dead space is, of course, equal to the anatomical dead space.

Instruments can also contribute to dead space, when the instrument is connected to the trachea by a hose or tube in which gas moves back and forth. This is undesirable. Consider what would happen when you would breathe through a long vacuum cleaner hose, whose volume is larger than your vital capacity. Due to the large volume of the hose, the oxygen-poor and carbon dioxide-rich gas that is exhaled in one breath remains in the tube and will be re-inhaled in the next breath; the exhaled carbon dioxide cannot be purged to the outside air, and the outside air cannot replenish the consumed oxygen, however strenuous the breathing.

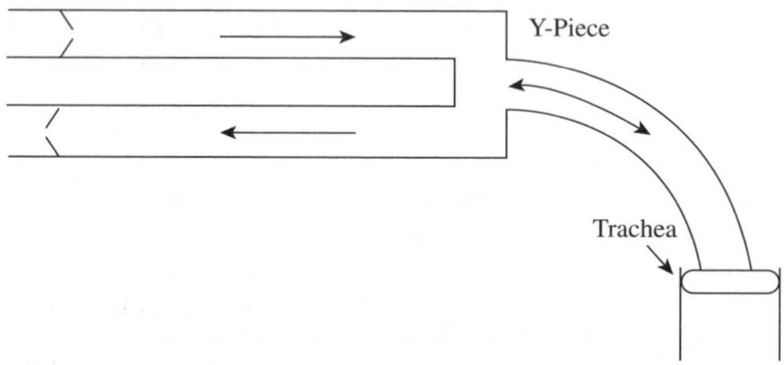

Figure 1.15. In a breathing system, device dead space is minimized by transporting inspiratory and expiratory gases through separate tubes.

For this reason, the dead space volume of connecting tubes or hoses must be as small as possible. A small volume implies a short and/or narrow tube. But a short tube is impractical and a narrow tube would normally introduce too much resistance to the gas flow (see Section 2.3). A good solution exists; it is based on our recognition that dead space exists only in tubes with a back-and-forth flow. If *two* tubes are used, an inspiratory and an expiratory tube, and if unidirectional valves ensure that the flow in these tubes is unidirectional, the tubes will *not* add to the dead space volume. The two tubes come together in what is called a Y-piece or T-piece (Figure 1.15), close to the mouth of the patient. The volume of the common leg of the Y-piece, in which the flow is bidirectional, can be made sufficiently small.

Question: To show the effect of instrument dead space, assume that the patient breathes through an extra piece of tubing of length 10 cm and diameter 3 cm. Calculate the volume of this tube. Assuming a normal adult dead space value of 150 ml, what percentage have we added to it? And assuming a neonatal dead space value of 20 ml, what percentage have we added?

1.2.8. Gas exchange

Atmospheric air mainly consists of nitrogen (about 79%) and oxygen (about 21%) with trace amounts of other gases. Expired gas contains about 16% oxygen (there is *oxygen consumption*) and 5% carbon dioxide (*carbon dioxide production*). Thus, only about one quarter of the oxygen that we inspire is taken up by the blood, and an approximately equal volume is returned as carbon dioxide. Another difference between the inspired and expired gases is that the latter is fully saturated with water vapor, whereas the former usually contains only little of it.

Fick's First Law states that the diffusion flow F of some medium (e.g., oxygen) through some surface with thickness x (e.g., the lung membrane) is proportional to the gradient of the concentration C through the surface

$$F = D\, dC/dx \tag{1.17}$$

where the proportionality constant D is called the *diffusion coefficient* for a unit surface (1 m^2) and the *diffusion capacity* if the surface area is unknown. Diffusion through the surface takes place from a high to a low concentration.

Instead of concentrations, we can also consider partial pressures. Since the total lung surface area is unknown, the parameter that characterizes the transport of a gas through the lung membrane is its diffusion capacity D, now redefined as the mass flow rate of that gas through the membrane divided by the partial pressure difference of the gas across the membrane.

$$D = F / \Delta P \tag{1.18}$$

The diffusion capacity has the character of a conductance (the inverse of a resistance) to the flow; note the similarity to Ohm's law.

If, at a certain flow F, the partial pressure difference across the membrane ΔP is small, the membrane's resistance to the gas flow is small and the gas can easily cross the membrane. The diffusion capacity is proportional to the alveolar membrane's surface area and inversely proportional to its thickness. It is also proportional to the gas solubility and inversely proportional to the square root of the molecular weight of the gas. Some gases may therefore pass the same membrane easily (oxygen, carbon dioxide, carbon monoxide); other gases may pass only with difficulty (nitrogen) or hardly at all (helium).

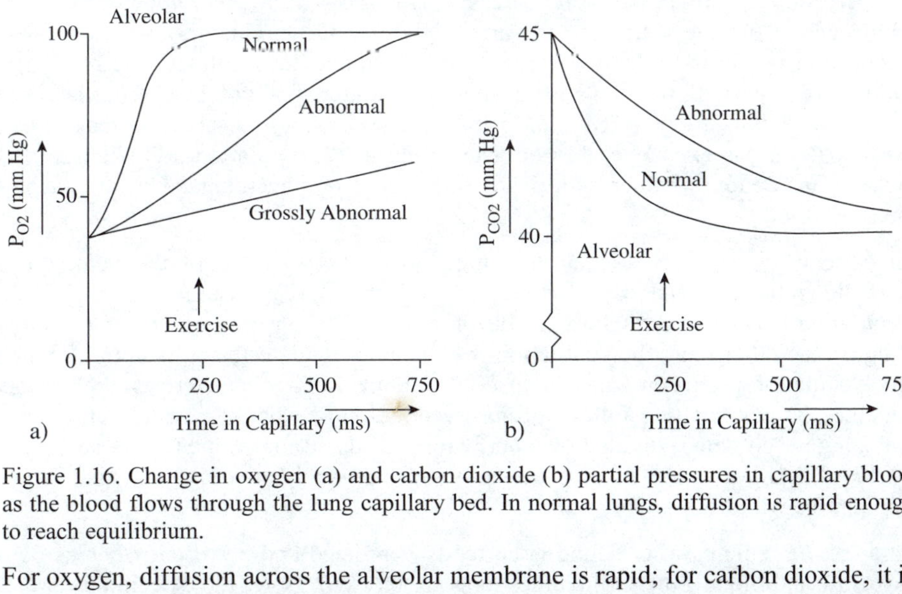

Figure 1.16. Change in oxygen (a) and carbon dioxide (b) partial pressures in capillary blood as the blood flows through the lung capillary bed. In normal lungs, diffusion is rapid enough to reach equilibrium.

For oxygen, diffusion across the alveolar membrane is rapid; for carbon dioxide, it is even about 20 times faster (Figure 1.16). In rest, red blood cells remain in the lung capillaries for about 0.75 s. During exercise, the cardiac output is larger and the time available for oxygenation is reduced. As long as exercise is moderate, however, diffusion of both O_2 and CO_2 is essentially complete in the normal lung.

Nitrous oxide (N_2O), an anesthetic, is a relatively insoluble gas with a large diffusion capacity. During surgery, the inspiratory gas contains about 70% N_2O; as a result, the N_2O partial pressure in the blood is high. When surgery is finished and the use of N_2O discontinued, it will rapidly leave the blood and fill up the alveoli. This temporarily reduces the alveolar volume that can be filled with inspiratory gas, which may lead to "diffusion hypoxia." For this reason, 100% O_2 is given for some time after the use of N_2O is discontinued.

1.2.9. The ventilation-perfusion ratio

Inadequate gas exchange between alveoli and pulmonary capillaries is caused by either a thickening of the alveolar wall, which causes a decrease of the lung's diffusion capacity, or by a decrease of the surface area over which gas exchange takes place. The latter normally means that some of the total alveolar surface area is not *ventilated* or that some of the area is not *perfused*.

In the expression *ventilation-perfusion ratio*, (alveolar) ventilation stands for the volume of gas that is transported into and out of the lungs each minute, and (alveolar) perfusion stands for the volume of blood that perfuses the alveoli per minute. A normal value of this ratio is between 0.8 and 1. The value of the ventilation-perfusion ratio is often used as a measure for the adequacy of gas transport across the lung membrane. If a part of the lung is blocked to ventilation but the perfusion is normal, the ratio decreases. This is a sign of *shunt flow*: some of the blood goes from the venous to the arterial side of the circulatory system without acquiring oxygen and releasing carbon dioxide. Shunt flow "dilutes" the blood gas values. If a part of the lung is normally ventilated but not perfused, the ratio increases. This is a sign of *physiological dead space*: parts of the lung that should participate in gas exchange, do not. The value of the ventilation-perfusion ratio is thus an indication whether significant shunt flow or physiological dead space is present.

In extreme cases such as the following, however, the value of the ventilation-perfusion ratio may lead to a false conclusion (Figure 1.17). Assume one lung whose ventilation is normal, but whose perfusion is zero. There is no gas exchange in this lung. In the other lung, the ventilation is blocked but the perfusion is normal. Since the ventilation-perfusion ratio is just one number that characterizes both lungs together, its computation follows from a ventilation of half the normal value (only one lung is ventilated) divided by a perfusion half its normal value (only one lung is perfused). Although the ratio is perfectly normal, we have a situation that is not compatible with life.

In a less extreme case, one lung is correctly ventilated and perfused, whereas both gas flow and blood flow to the other lung are blocked. Here, too, we find a normal ventilation-perfusion value. This is not a life-threatening situation (people do survive after removal of one lung), but it is clear that the ventilation-perfusion ratio does not tell the whole story and must be used in combination with other information.

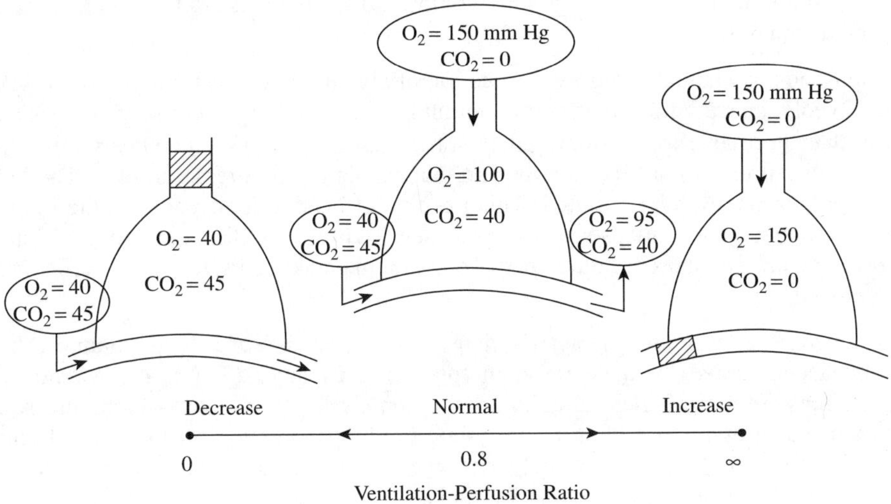

Figure 1.17. The ventilation-perfusion ratio, which is normally 0.8, decreases if parts of the lung are not ventilated and increases if parts of the lung are not perfused.

Question: How does the ventilation-perfusion ratio change if a patient is accidentally ventilated by a tube into one of the two main bronchi rather than in the trachea?

Question: How does the ventilation-perfusion ratio change if a patient is accidentally ventilated by a tube into the esophagus rather than in the trachea?

Question: How does the ventilation-perfusion ratio change in case of pulmonary embolism (a full or partial obstruction of the flow in the pulmonary artery)?

1.2.10. Some pathophysiologies

Pathologies of the lung mechanics can be classified as either *obstructive* or *restrictive*.

Obstructive diseases are characterized by an increase of the airway resistance, which can be found in, e.g., asthma, emphysema, chronic bronchitis or pulmonary fibrosis. Obstructions make it difficult to inhale and exhale. Patients with obstructive diseases require much effort just to breathe.

Asthma is an allergy of the airways, which results in inflammation and an increase in tone of the bronchial smooth muscles that control the diameter of the airways. In an acute asthma attack, bronchospasm decreases the diameters of the airways even more. Due to the smooth muscle contractions, gas may become trapped in parts of the lung behind closed airways. Extra respiratory effort is required to transport the respiratory gas through the extra-narrow passages, and the resulting extra-high intra-thoracic pressure may close off collapsible airway passages and thus lead to exaggerated airway collapse (see Section 2.9). To minimize collapse, a patient with acute asthma will breathe at high lung volumes, inspiring close to total lung capacity (see Section 6.3).

Emphysema is characterized by destruction of alveolar walls, resulting in dilation of the alveolar spaces and a very compliant lung. The high lung compliance means that less than normal effort is required to inhale ($\Delta P_{muscle} = \Delta V/C$), but also that the expiratory time constant RC is large and that exhalation progresses slowly. The lung is kept chronically hyperinflated. This has two effects: at high volumes the lung is less compliant and collapse is lessened (see Figure 1.6). The high lung volume, however, flattens the curvature of the diaphragm, making its force less efficiently employed.

Restrictive diseases are characterized by a decrease in useful lung volume (VC), often accompanied by a decrease in total lung capacity (TLC). An example is pulmonary fibrosis, where the lungs are abnormally stiff. Very large drops in intrapleural pressure are required to inflate the lungs, so deep breaths are difficult. Patients tend to breathe shallowly and rapidly. Restrictions can also be due to rigidity of the chest wall or to muscle weakness or paralysis.

1.3. ANATOMY AND PHYSIOLOGY OF THE HEART AND THE CIRCULATORY SYSTEM

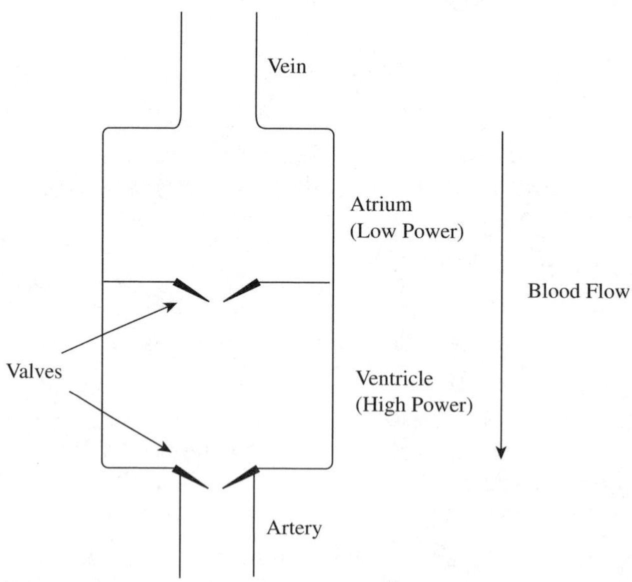

Figure 1.18. The valves of the heart ensure that blood flows in one direction.

Anatomical analysis of the circulatory system shows that it consists of two synchronously acting pumps called the left and the right[1] heart, two highly branching *distributing* networks of vessels called the pulmonary and the systemic arterial systems and two highly branching *collecting* networks of vessels called the pulmonary and the systemic venous systems. In between arteries and veins we find a very large number of tiny peripheral vessels that bring the blood into close proximity of all the cells.

Each of the two pumps, the left and the right heart, consists of two chambers, called the atrium and the ventricle (Figure 1.18). The ventricles are high power pumps that, when they contract, generate enough pressure to force the blood through the small peripheral blood vessels. The atria have much less power; it is their function to pump blood into the ventricles, where the real work is done.

Between each atrium and the corresponding ventricle we find a one-way valve, which prevents blood from flowing backward. Valves are also found where the blood leaves the ventricles. In contrast with the respiratory system where gas flow is back-and-forth, the valves of the circulatory system force blood to flow unidirectionally, in a "circle" (see Figure 1.1). Note that there is no valve where the blood enters the atrium; such a valve is not necessary because when the atrium contracts, the ventricle is fully relaxed and, through an easily opened valve between atrium and ventricle, is easily filled with blood.

1.3.1. Blood transport

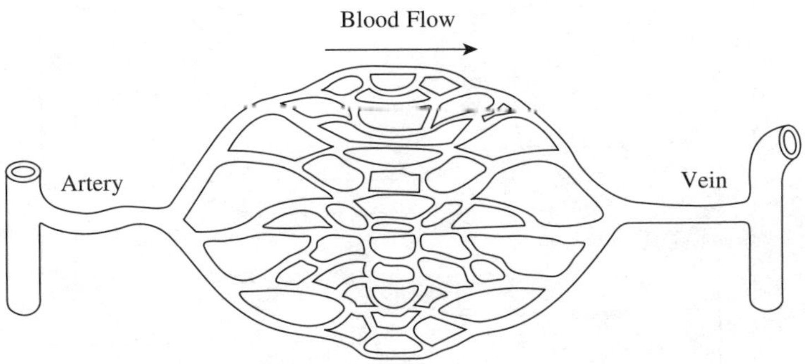

Figure 1.19. Arteries branch into successively smaller vessels. The smallest vessels are the capillaries. The vein collects the blood again.

The right heart pumps blood from the systemic veins through the pulmonary circulation toward the left heart, where an adequate filling pressure for the left atrium must be provided. The left heart pumps the blood from the pulmonary circulation into the aorta, from where it is further distributed to the organ tissues.

[1] Right and left are from the patient's, not the observer's perspective. Drawings, however, show the point of view of an observer looking at the patient's front.

The parts of the lung where no gas exchange takes place (the conductive zone) and the heart itself are also "organs" in this sense, which must be provided with oxygenated blood.

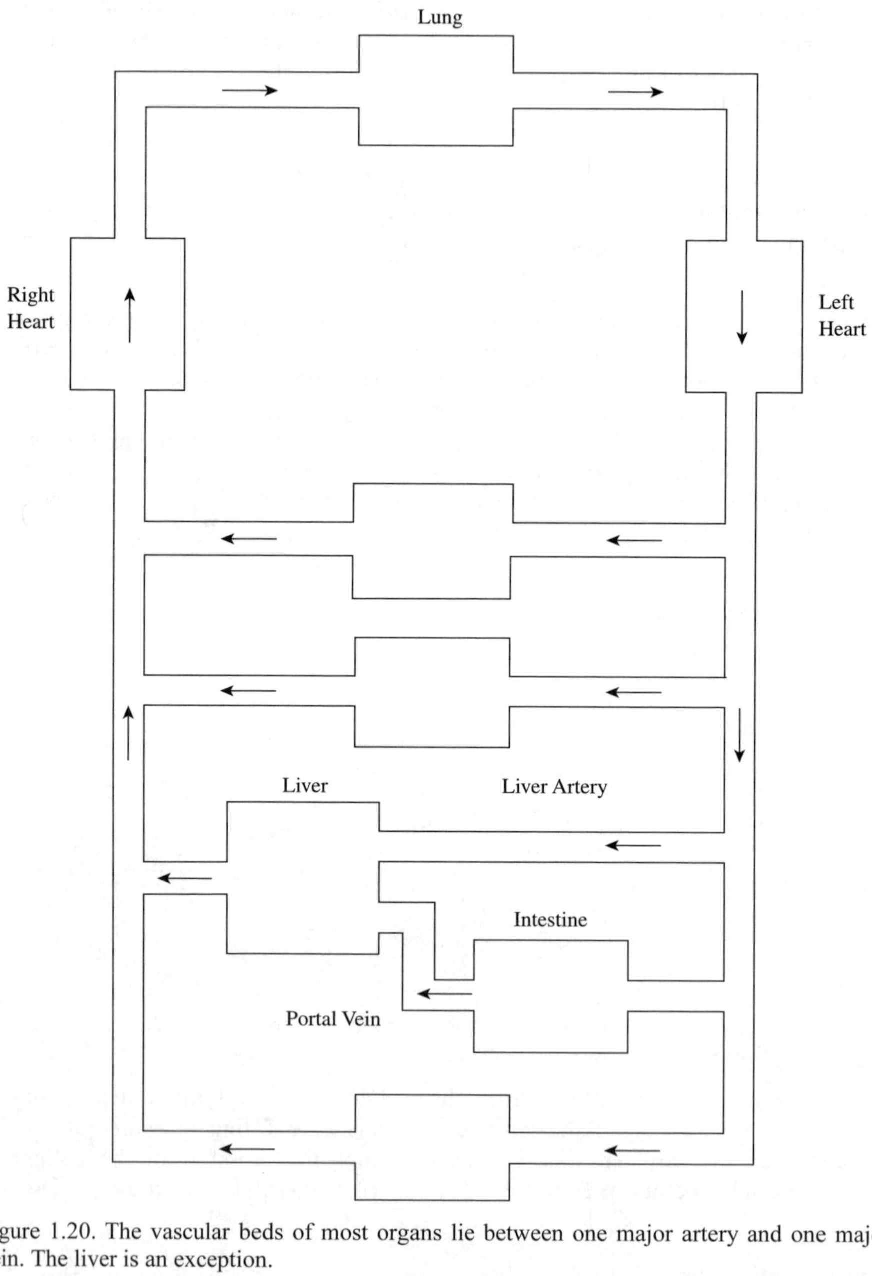

Figure 1.20. The vascular beds of most organs lie between one major artery and one major vein. The liver is an exception.

The volume of blood ejected by each cardiac contraction is called the *stroke volume* (70 ml at rest). At rest the *heart rate* is about 70 beats per minute. The blood flow pumped by the heart is thus about 70 ml times 70/min = 5 l/min at rest (of which 20% goes to the muscles), rising to 25 l/min during strenuous exercise (of which now about 85% goes to the muscles). This "volume-per-minute" (flow rate) is called the *cardiac output*. Since the total blood volume is about 5 l, the average *circulation time* of the blood is about 1 min. The right and the left heart contract at the same times and in the same rhythm, so the stroke volumes of left and right heart must be the same on average.

The large arteries that leave the ventricles successively branch into ever smaller vessels (Figure 1.19), much like in the respiratory system. Gas exchange takes place in the smallest vessels, the *capillaries*. Another branching network at the venous side returns the blood to a vein. Thus, the blood that is provided by one artery is usually collected by a single vein: each organ has its own *vascular bed*. This is a general rule: each organ's blood supply is maintained by *one* major artery and *one* major vein. The liver, however, has *two* major blood supplies, the liver artery and the portal vein (Figure 1.20).

Question: Can you think of a reason why the intestine and the liver are connected in series? Hint: what is the function of the liver?

The *pulmonary arterial system* transports the blood to the lung. It also buffers the pulsatile pressure output of the right heart to a more constant pressure. The pulmonary peripheral circulation consists of those small blood vessels that surround the alveoli of the lung; this is where oxygen is picked up by the blood and carbon dioxide released. The pulmonary venous system collects the oxygen-enriched blood from the pulmonary peripheral circulation, returning it to the left heart.

The *systemic arterial system* distributes the oxygenated blood to the various organs; it also buffers the pulsatile pressure of the left heart to a more constant pressure. The systemic peripheral circulation provides oxygen to the interstitial fluid surrounding all cells and collects excess carbon dioxide. The systemic venous system collects the oxygen-poor blood that has traversed the organs to return it to the right heart and the lung for renewed oxygenation. It also functions as a pool where variations in total blood volume are taken up. The volume of blood at the venous side of the systemic circulation (3000 ml) is much larger than that of the arterial side (750 ml), and the compliance of the veins is much larger than that of the arteries. Rapidly adding a blood volume of 100 ml to the veins causes a pressure rise of 1 mm Hg, whereas the same pressure rise in the arteries is obtained by adding only 2.5 ml. This means that the veins can store large volumes of blood at low pressures. Table 1.2 shows the major differences between the systemic and the pulmonary circulation.

Although the above description of the circulation might give the impression that the circulation is a closed system, it is not — and it *should* not be closed. In fact, the endothelial cells, which line the capillaries, normally block only the passage of blood cells and large protein molecules. Fluids and small molecules freely pass between blood and the interstitial space. Moreover, fluid is added to the circulation from the intestines and removed from the circulation — together with superfluous electrolytes and waste products — via the kidneys. Yet, although the circulation is essentially open for fluids, a number of control processes operate to stabilize the filling of the circulatory system to about 5 or 6 l in adults. When these control systems break down, "shock" (too small circulating volume) or hypertension (too large circulating volume) may result. Shock is especially threatening; at an arterial pressure lower than about 50 mm Hg the oxygen supply to the brain is inadequate. The subject faints and must be helped immediately. If the shock persists for longer than a few minutes, it becomes lethal.

Table 1.2. Differences between the systemic and the pulmonary circulation.

systemic circulation	pulmonary circulation
feeds many organs	feeds the lung only
adapts to different demands	one function only
many control systems	few control systems
large a-v pressure difference	small a-v pressure difference
large flow resistance	small flow resistance
long vascular bed (extremities)	short vascular bed (lungs only)

A great variety of additional (local and central) circulatory control processes operate as well. We mention only a few. Regulation of the regional blood flow to the various organs tunes itself to the oxygen requirements of those organs through the contraction of the muscles that control the diameters of the arterioles which feed the organs. Thus, at different times different volumes of blood will perfuse the organs, depending upon the "work" the organs have to do. The venous volumes adapt to variations in the flows of blood offered to them, thus keeping pressures within narrow bounds. Other control systems regulate arterial blood pressures, protecting the vessel walls from stressfully high tensions, yet ensuring adequate perfusion of the various tissues. Another set of control systems (see Section 1.2.9) harmonizes respiration and circulation. When these break down, one speaks of ventilation-perfusion mismatch.

1.3.2. Blood; blood gases; hemoglobin; hematocrit

Blood is a fluid — this fluid is called *plasma* — in which we find three types of free-floating cells: red and white blood cells and platelets. The plasma also contains ions (such as Na^+, K^+ and Cl^-) and a variety of proteins (about 7% of the plasma's mass). The cells make up about 45% of the blood volume; this percentage is called the blood's *hematocrit*. The major function of platelets (also called *thrombocytes*) is in the repair of damaged blood vessels, where they direct fibrinogen, one of the proteins that is found in the blood, to form a mesh of fibers that repairs the wound. The major function of the white cells (also called *leukocytes*) is to find and destroy foreign material that has made its way into the blood or the tissues.

The red blood cells, also called *erythrocytes*, are the most numerous. In fact, fully one-third of all the cells of the human body are erythrocytes. Their major function is oxygen transport. Hemoglobin (hgb), an iron-containing protein which makes up about 30% of an erytrocyte's volume, easily bonds with oxygen, forming oxyhemoglobin (hgbO). This is a so-called *equilibrium reaction*, because the reaction can proceed in either direction:

$$hgb + O_2 <=> hgbO \qquad (1.19)$$

When the oxygen partial pressure in the blood is high (especially in the blood perfusing the alveoli), oxyhemoglobin is formed. When the oxygen partial pressure in the blood is low (in the systemic peripheral circulation, where the cells consume oxygen, effectively removing it from the blood), oxyhemoglobin is converted back into hemoglobin. Figure 1.21 shows the relationship, which is known as the oxygen dissociation curve. This mechanism keeps the O_2 partial pressure in the peripheral blood almost constant (40 to 50 mm Hg), although in hard-working muscles it can drop to almost zero. At the normal systemic venous P_{O2} of 40 mm Hg, hemoglobin is 75% saturated. Thus, only 25% of the oxygen has dissociated from hemoglobin and entered the tissues. A rough calculation shows that after 4 passages through the tissues, all oxygen would have been consumed if oxygen uptake from the lung would stop. Since the blood circulates in about 1 min, this indicates an oxygen reserve in the order of 4 min. The figure also indicates that raising the P_{O2} above 80 or 100 mm Hg will hardly increase the *oxygen content* (dissolved plus bound oxygen) of the blood.

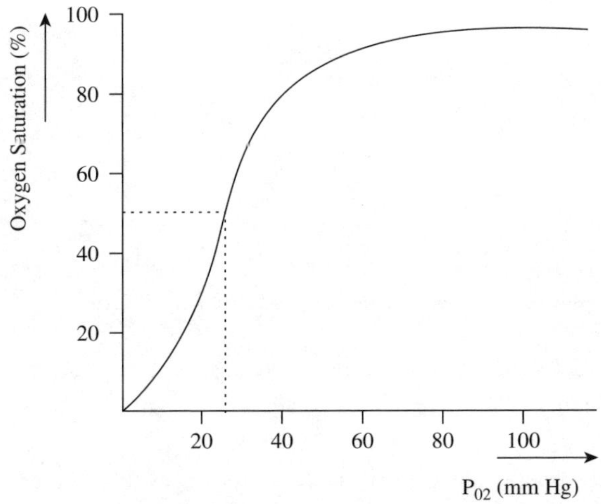

Figure 1.21. The oxygen dissociation curve. If the oxygen partial pressure is above some 80 mm Hg, the saturation is essentially 100%.

Question: Figure 1.21 indicates that a P_{O2} of more than 80 mm Hg does not lead to a significantly higher oxygen saturation. Thus, ventilating a patient with a gas containing a high oxygen percentage normally makes no sense. Which percentage would normally suffice?

The oxygen buffering by the oxyhemoglobin is essential. One liter of blood plasma (without hgb) can contain only 3 ml of O_2. One liter of whole blood, with hemoglobin, can contain about 200 ml O_2. More than 98% of the oxygen in the blood is thus bound to hemoglobin. Oxygen supply to the cells would be far from adequate through only dissolved oxygen.

Some of the hemoglobin can be bound to carbon monoxide in the form of carboxyhemoglobin (hgbCO), especially in smokers. Since hgb binds 200 times more easily with CO than with O_2, CO will even replace the hgbO's O_2. Moreover, hgbCO is rather stable. As a result, smokers have less hgb available to transport oxygen.

The hematocrit determines the blood's oxygen-carrying capacity. The higher the hematocrit, the more hemoglobin and the more oxygen can be carried; this relationship is approximately linear (Figure 1.22):

$$TO_2 = K_1 \text{ Hct Sat F} \tag{1.20}$$

where TO_2 is the amount of oxygen carried per time unit, K_1 is a constant, Hct the blood's hematocrit value (which is assumed to be linearly related to hemoglobin concentration), Sat the blood's oxygen saturation and F the blood flow through the vascular bed. This means that the oxygen transport capacity is *higher* with a high hematocrit.

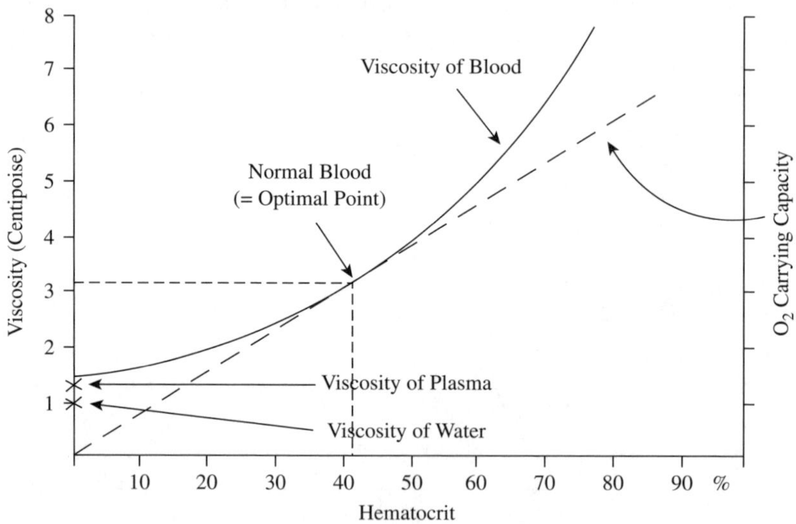

Figure 1.22. Blood viscosity and maximum oxygen content as functions of hematocrit.

The hematocrit also determines the blood's viscosity η. A high hematocrit means a high viscosity; this relationship is approximately quadratic:

$$\eta = \eta_0 (1 + K_2 \text{ Hct}^2) = \eta_0 + K_3 \text{ Hct}^2 \tag{1.21}$$

where K_2 and K_3 are constants.

Since a higher blood viscosity means an increase in flow resistance (see Section 2.2)

$$R = K_4 \eta \quad (1.22)$$

(where K_4 is another constant), the blood flow F through the vascular bed will decrease — if the pressure P remains the same — as viscosity rises:

$$F = P / R \quad (1.23)$$

Since the blood flow F carries the oxygen, this means that the oxygen transport capacity is *lower* with a high hematocrit. In reality, there is an optimum hematocrit value, which ensures best oxygen transport — and this optimum value is the value that we measure in healthy persons. Thus, athletes who use "blood doping" actually *decrease* their performance.

Using the above relationships, we find

$$TO_2 = \frac{K_1 PSat}{K_4} \frac{Hct}{\eta_0 + K_3 Hct^2} \quad (1.24)$$

Differentiating this expression with respect to Hct and equating to zero gives us the optimal value of the viscosity (η_0 = 1.6 cPoise[1] gives η_{opt} = 3 cPoise) and of the hematocrit (about 40%):

$$Hct^2 = \eta_0 / K_3 \quad (1.25)$$

Blood carbon dioxide is also buffered. Although erythrocytes buffer some CO_2, the major mechanism is very different. Carbon dioxide and water form carbonic acid, which partly ionizes

$$CO_2 + H_2O <=> H_2CO_3 <=> H^+ + HCO_3^- \quad (1.26)$$

These are equilibrium reactions as well. If the CO_2 partial pressure in the blood rises, the reaction proceeds from left to right, producing more carbonic acid, which splits into a hydrogen ion and a bicarbonate ion. When the hydrogen ion concentration rises, the reaction proceeds to the left, resulting in the production of CO_2. This mechanism keeps the CO_2 partial pressure in the blood almost constant. In systemic arterial blood, about 90% of the carbon dioxide is in the form of bicarbonate. The same mechanism also controls the body's *acid-base balance* (the acidity of the blood, i.e., its pH, is determined by the H^+ concentration). Both too acidic (low pH) and too alkaline (high pH) cell environments are prevented by the above equilibrium reaction.

Several other chemical reactions modify these basic oxygen and carbon dioxide dissociation processes. Their combined effect is the extra facilitation of oxygen uptake and carbon dioxide release in the lung and oxygen release and carbon dioxide uptake in the tissues.

In addition to these "passive" chemical reactions, internal control systems adjust respiration in accordance with the needs of the tissues. Although minute volume is increased by a decrease in arterial P_{O2}, this only happens when the decrease is large.

[1] The centipoise is a unit often used. 1 Poise = 0.1 Pa.s; 1 cPoise = 0.001 Pa.s.

Even a slight increase of arterial P_{CO2}, however, is rapidly compensated for. The stimulus for this reflex is not the increased P_{CO2} itself, but the concomitant increase in H^+ concentration in the arterial blood and in the brain. During moderate exercise, blood gas concentrations and pH remain unchanged, because ventilation increases in exact proportion to metabolism. During very strenuous exercise, full compensation is no longer possible.

1.3.3. Volume-pressure relationships of heart and blood vessels

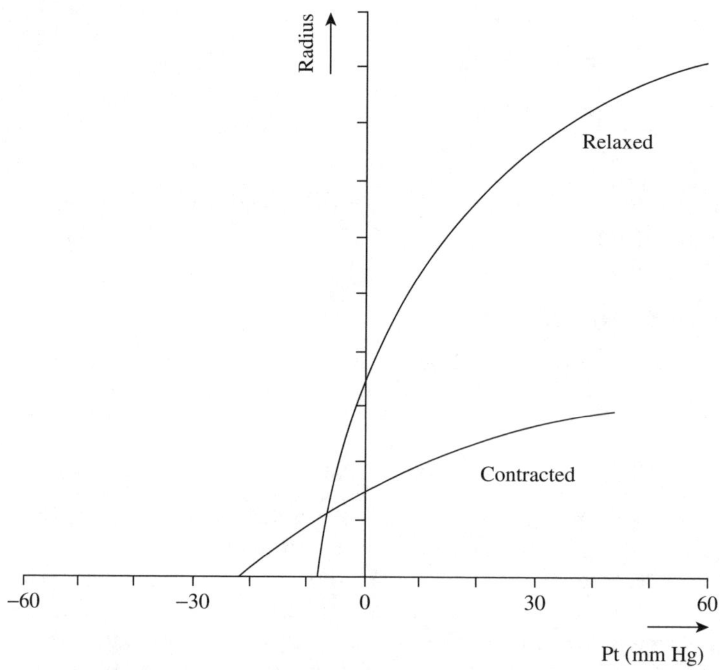

Figure 1.23. Transmural pressure-diameter curve of a typical artery. The diameter is in arbitrary units. The slope of this curve is related to the compliance: at higher pressures the artery is stiffer.

The main difference between arteries and veins is in their compliances and their radii. Figure 1.23 shows the pressure-diameter relationship of a typical blood vessel, from which a volume-pressure relationship can be computed. As with the vessels of the respiratory system, we notice that a straight-line approximation works well over most of the normal operating range. The slope of the line determines the compliance, and the intersection of the line with the volume axis determines the unstressed volume. Notice that the compliance becomes smaller — the vessel becomes stiffer — at high filling volumes. Table 1.3 shows some representative values. It shows that veins are much more compliant than arteries and that they have much larger unstressed volumes.[1]

That unstressed volumes are not zero but relatively large has an important consequence. The total blood volume is about 5 l. An abrupt 1-l decrease of the volume does not cause a pressure drop to 80% but to 30% of the original pressure. A volume increase of only 1.6 l builds up all of the pressure in the system; a loss of this 1.6 l volume results in zero pressure and thus no perfusion. The remaining volume of 3.5 l fills the unstressed volume. A slower decrease of the blood volume, e.g., for transfusion, does not cause noticeable pressure drops, due to the physiological control systems that stabilize the blood pressure.

Table 1.3. Some properties of various blood vessels. Stressed volume can be computed as the difference between total volume and unstressed volume. The compliance values are at normal (total) volumes.

	transmural pressure (mm Hg)	total volume (ml)	unstressed volume (ml)	compliance (ml/mm Hg)
intrathoracic arteries	100	250	180	0.7
extrathoracic arteries	100	500	320	1.8
capillaries	20	200		
extrathoracic veins	4	1500	1250	60.0
intrathoracic veins	8	1500	1000	60.0
lung arteries	15	150	85	4.3
lung capillaries	10	100		
lung veins	8	600	400	26.0

1.3.4. Pressures throughout the circulation

Pressure variations in the heart's ventricles are large. The heart receives its inflow from the veins, where pressures are as low as 5 mm Hg, and pumps the blood to the arteries, where (systemic) mean pressures can be as high as 200 mm Hg. Pressure drops in the large vessels are low due to their large diameters; the pressure drop occurs mainly (about 80%) in the small-diameter peripheral arterioles. Figure 1.24 demonstrates this for both the systemic and the pulmonary circulation. The latter is smaller in extent, its outflow pressure is lower (about 15 to 30 mm Hg) and its pressure fall at the capillaries is less abrupt. The differences between the pressure levels in the pulmonary and the systemic circulation are due to a number of reasons: the pulmonary capillaries are wider than the systemic ones; the alveolar capillaries have no smooth muscles that can control their diameter, which causes a very constant ratio between pressure and flow; Thus, the resistance of the alveolar

[1] To a large extent, this may be due to the different environmental conditions (e.g., pressures) to which arteries and veins are exposed. In coronary artery bypass surgery, for instance, faulty arteries are routinely — and successfully — replaced by veins excised from a leg of the patient.

capillary bed is rather small (one eighth of that of the systemic circulation) and constant; the total capillary volume is relatively large, and so is its reserve volume (it can expand well).

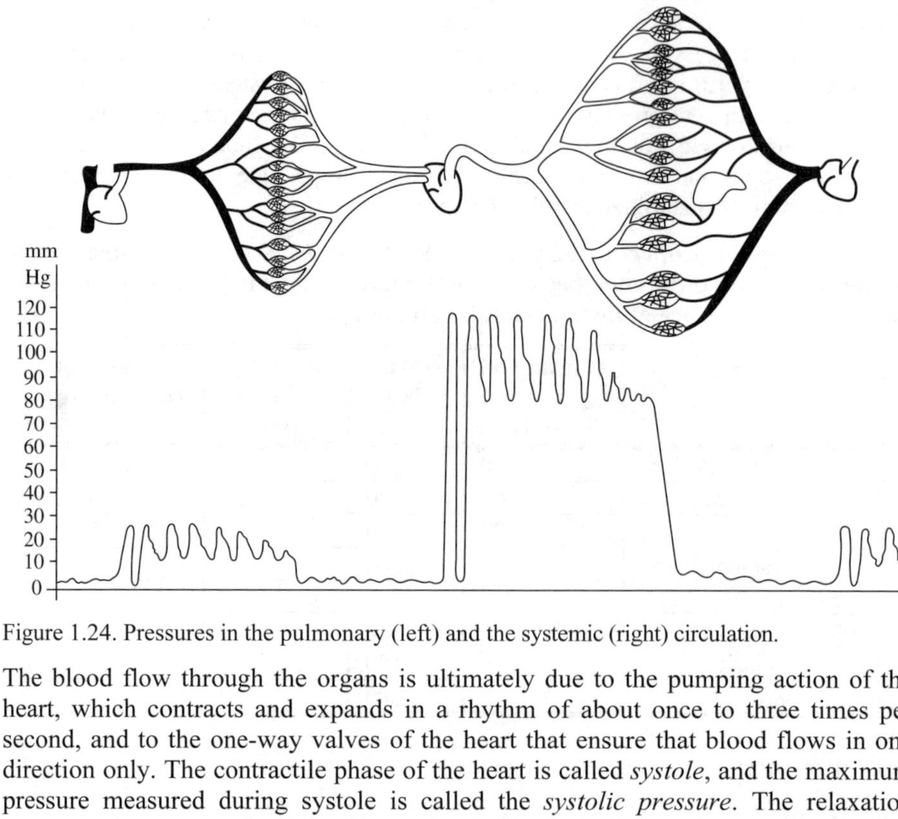

Figure 1.24. Pressures in the pulmonary (left) and the systemic (right) circulation.

The blood flow through the organs is ultimately due to the pumping action of the heart, which contracts and expands in a rhythm of about once to three times per second, and to the one-way valves of the heart that ensure that blood flows in one direction only. The contractile phase of the heart is called *systole*, and the maximum pressure measured during systole is called the *systolic pressure*. The relaxation phase of the heart is called *diastole*, and the minimum pressure measured during diastole is called the *diastolic pressure*.

The blood flow through the organs is controlled by the smooth musculature in the walls of the arterioles, which controls their cross section (lumen[1]) according to the oxygen demands of the organs. The small radius of the capillaries means that they can be modeled as a pure resistance (see Section 2.5).

The term *peripheral resistance* is used to indicate the total resistance that the flow pumped by the heart meets on its way from ventricle to atrium. During the time period when the heart valve between the ventricle and aorta is open, the arterial pressure can, in a first approximation, be modeled as the product of peripheral resistance and the outflow of the ventricle

[1] The word lumen is used in two different meanings. Quantitatively, the lumen of a tube or vessel is its inside cross-sectional area, expressed in, e.g., cm^2. Qualitatively, it can also mean the three-dimensional hollow interior "tube" within the physical tube.

$$P_{art}(t) = F(t) R_{peri} \tag{1.27}$$

when we neglect the small pressure drop over the aortic valve and the arterial compliance. The peak arterial pressure, the *systolic arterial pressure*, is then due to the maximum outflow that the heart can realize and the value of the peripheral resistance (Figure 1.25).

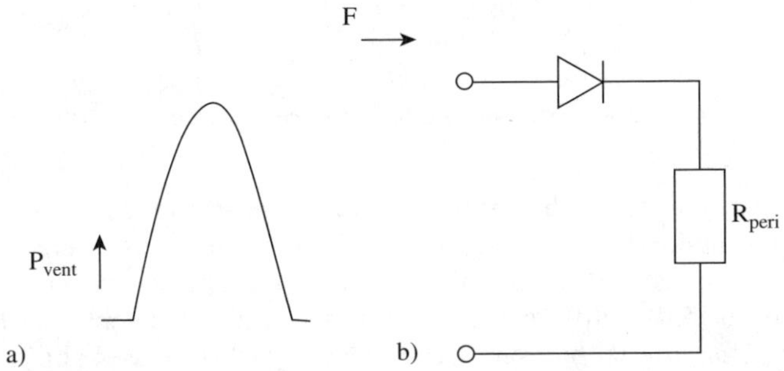

Figure 1.25. Model of the systemic circulation if the arterial compliance is neglected.

This approximation is valid only if the arterial compliance is very low, such as in severe atherosclerosis, or can be neglected, as when a heart-lung machine generates a constant flow F. Since normally we cannot neglect the arterial compliance, the above model needs to be modified: the systolic arterial pressure is due to the maximum outflow that the heart can realize and the values of peripheral resistance and arterial compliance (Figure 1.26). Since the heart generates a flow into the aorta rather than a pressure (see Section 1.3.5), the effect of a small arterial compliance is large: severely atherosclerotic patients may have extremely elevated systolic pressures.

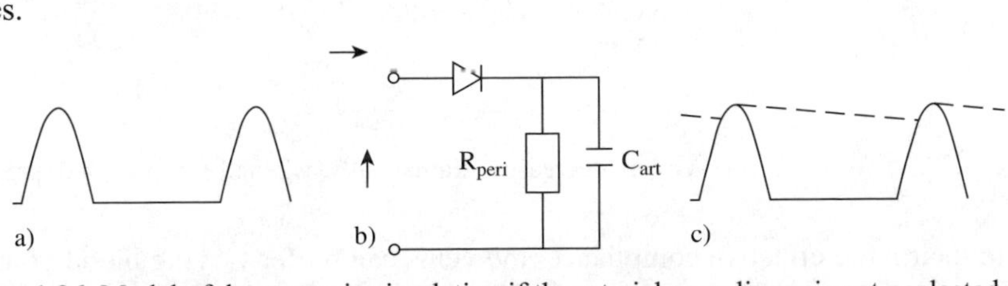

Figure 1.26. Model of the systemic circulation if the arterial compliance is not neglected.

Due to the aortic valve, no blood can flow back from aorta to ventricle; during diastole, the aortic pressure is thus isolated from the ventricular pressure. Thus, the minimum arterial pressure, the *diastolic arterial pressure*, does not directly correspond with the pressure in the ventricle. It depends on the arterial pressure at the end of systole (when the valves close), on the blood volume in and the compliance of the arteries, and on the duration of the diastole (Figure 1.26). The decay of the diastolic part of the arterial pressure is approximately exponential; in the model of Figure 1.26 it is the voltage of capacitor C_{art} when it discharges through the peripheral resistance R_{peri}.

The difference between systolic and diastolic pressure is called the *pulse pressure*. The duration of the systolic period is relatively constant; changes in heart rate, therefore, mainly have an effect on the diastolic period.

The average outflow of the heart is called the *cardiac output* (CO), and the average pressure in the systemic arterial pressure is called the *mean arterial pressure* (MAP). We therefore also have the following relationship

$$MAP = CO\ R_{peri} \tag{1.28}$$

Question: Calculate the value of the peripheral resistance, assuming that MAP and CO have normal values.

The arterial system vessels, and particularly the large arteries close to the heart, have very elastic (compliant) walls. They buffer the intermittently offered blood supply into a reasonably constant flow into the capillary bed. The *waveform* of the pressures that can be measured at different arterial sites is site dependent (see Figure 1.24), although the *mean* arterial pressure hardly changes (it decreases slightly toward the periphery). The waveform changes have two causes. The first is the frequency and phase distortions that arise when the pressure pulse travels down the artery. We will analyze this in more detail in Section 3.2.1.1. The second is the reflections of the pressure pulse that arise wherever the vessel radius changes or the vessel branches.

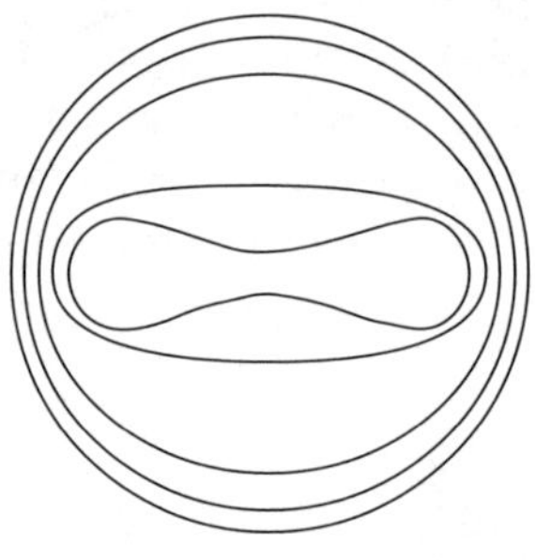

Figure 1.27. A vein's cross section decreases dramatically when the transmural pressure increases.

Due to their large effective compliance, the veins can buffer varying blood volumes at approximately constant pressures. Pressure variations in veins are thus small.[1] Fluid flow in the venous system is also different from flow in the arteries. In veins, internal pressures are low, and transmural pressures are thus significantly influenced by small changes in outside pressures. Pressure variations outside the vein are caused by breathing or ventilation and by the contractions of various muscles. Since arteries frequently share a common inelastic sheath with veins in which they run adjacent, arterial pulsations also cause volume variations of veins. The result of an outside pressure which is higher than the pressure inside the vein may be gradual or

[1] That is, pressure variations due to the pumping of the heart. Pressures in the veins may be comparable to those in the arterial system, especially in the veins of the lower leg in an erect person. Pressure *variations* in these veins are also large when changing position from standing up to lying down or the other way around.

total venous collapse. The vein first loses its circular cross section and becomes elliptic, then becomes bi-elliptic, and finally may be closed off almost completely (see Figure 1.27). Veins also contain unidirectional valves in many places, mainly peripherally. In combination with venous collapse, these valves cause changes in transmural pressure to be effective in the transport of blood through the veins. All this makes it difficult to linearize (and to construct simple but realistic models of) the pressure-flow relationships in veins.

1.3.5. Cardiac mechanics

The pumping action of the heart causes time-varying pressures and volumes. Figure 1.28 shows left ventricular pressure, aortic pressure and left ventricular volume during a cardiac cycle.

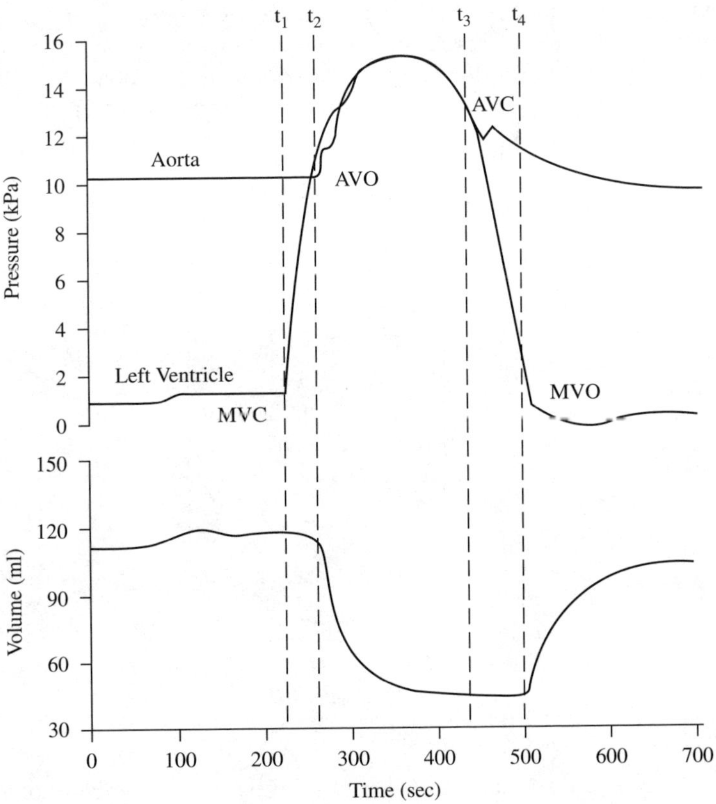

Figure 1.28. Left ventricular and aortic pressures and ventricular volume during one heart beat. At times t_1, t_2, t_3 and t_4, valves AV (aortic valve) and MV (mitral valve) open (O) and close (C). See text for more details.

At the end of diastole (before t_1), the ventricle is relaxed and its pressure is so low that it can be filled by atrial blood through the open mitral valve (the valve between left atrium and left ventricle). Ventricular contraction is initiated by the spread of electrical excitation across the ventricle, which is almost instantaneous; in normal adults, the duration of the QRS complex in the ECG is only about 60 ms (see Figure 1.31).

At t_1, systole (the contraction of the ventricle) starts and the mitral valve closes essentially immediately. During this phase, the ventricular pressure is still lower than the aortic pressure, so no blood can flow; the pressure rise is *isovolumetric* (occurs at an unvarying volume). At t_2, the pressure in the ventricle becomes greater than the aortic pressure and the aortic valve (the valve between left ventricle and aorta) opens. Blood starts to flow from ventricle to aorta. Then the ventricle relaxes again. At t_3, the blood flow stops, because the relaxation of the ventricle has progressed so far that the ventricular pressure has become equal to the aortic pressure. The pressure across the aortic valve becomes negative and it closes; the subsequent pressure drop is again *isovolumetric*. The small, rapid increase of the aortic pressure at t_3 (the *dicrotic notch*) is caused by the inertia of the blood. The blood, which flows into the aorta must decelerate to zero velocity. It even briefly reverses direction. This backflow actively closes the aortic valve. At t_4, ventricular relaxation is complete; the ventricular pressure becomes lower than the atrial pressure and the mitral valve opens again.

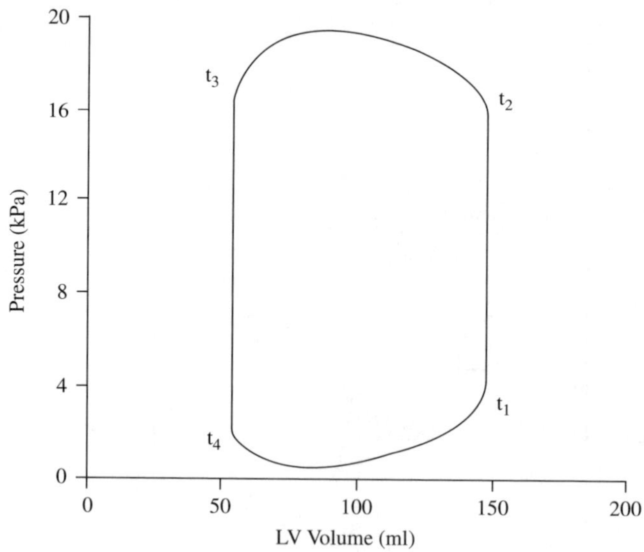

Figure 1.29. Left ventricular pressure-volume loop.

If we take the pressure and the volume values of Figure 1.28 at every time and plot these as points with pressure and volume coordinates, we obtain the ventricular *pressure-volume loop* of Figure 1.29. This transforms every *time* into a *point* on the loop. In Figure 1.28, the isovolumetric phases are those time periods where the volume does not change; in Figure 1.29, these are represented by straight vertical line segments. The points where valves open and close are the points where vertical line segments begin and end. The lower limb represents ventricular filling and the upper limb is the ejection phase. The volume change between the vertical iso-volumetric lines is the *stroke volume*; this same volume is called the *ejection fraction* when it is expressed as a fraction of the end-diastolic volume. The area enclosed by the curve is called *stroke work*, the mechanical work that the ventricle has to perform in order to eject its blood.

Question: Give a mathematical expression (or definition) for stroke work as a function of ventricular pressure and volume.

Ventricular volume is difficult to measure. Stroke work is therefore often approximated as the product of mean arterial pressure and stroke volume.

Question: Give an explanation why this is approximately correct. Would knowledge of the value of the central venous pressure improve the estimate? If so, how much?

Figure 1.30 shows several pressure-volume loops. Changes in the filling pressure of the ventricle (preload) move the end-diastolic point along a curve which is called the end-diastolic pressure-volume relation (EDPVR), which is mainly determined by the (non-linear) *passive* compliance of the (inactive) myocardium. The area under this passive curve is the work done by the blood on the ventricle during the ventricle's filling time. This work, which has to be performed by the left atrium, is small.

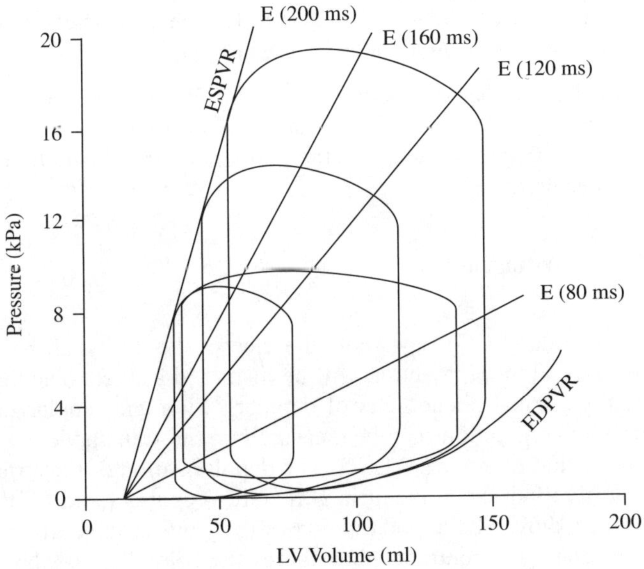

Figure 1.30. If several left ventricular pressure-volume loops are plotted, the ESPVR and the EDPVR are discovered. Both are, to a fair approximation, straight lines, particularly at low volumes.

Increases in ejection pressure (afterload) decrease stroke volume. The end-systolic pressure-volume relation (ESPVR) connects the end-ejection points, where the aortic valve closes. It is approximately linear and almost independent of stroke volume.

Figure 1.30 also shows that stroke work increases with ventricular filling; this is called the Frank-Starling mechanism or Starling's "law of the heart." This mechanism results from the properties of the myocardium: the energy of contraction is a function of the length of the muscle fibers. Starling's law more simply stated: the ventricle ejects whatever volume of blood was put into it.[1] The greater the filling of the ventricle, the stronger the subsequent contraction.

Connecting pressure-volume points at corresponding times in the cardiac cycle also results in almost straight lines throughout systole; the intercepts of these lines with the volume axis approximately come together in a single point V_0. The ventricular pressure P (t) thus is roughly proportional to the ventricular volume V (t) through a time-varying elastance E (t)

$$P(t) = E(t)\,(V(t) - V_0) \qquad (1.29)$$

This time-varying elastance $E = \Delta P/\Delta V$ has the character of a (time-varying) compliance,[2] defined as $C = \Delta V/\Delta P$. But whereas a compliance is due to the *passive* characteristics of the tissue, this elastance has an *active* character: it is due to the contraction of the cardiac muscle. The maximum elastance E_{max} (the slope of ESPVR), also called the heart's *contractility*, is a preload- and afterload-independent indicator of how powerful the heart can contract. Drugs, e.g., catecholamine, can increase E_{max}, whereas E_{max} shows a decrease in less healthy hearts. Contractility, as well as heart rate, can also be modulated by the nervous system.

Although the stroke volumes of left and right ventricle are identical, left ventricular stroke work is approximately seven times larger than right ventricular stroke work. The reason is simply that the pressure in the pulmonary artery is only one-seventh of the pressure in the aorta.

1.3.6. The electrocardiogram

The electrocardiogram (ECG) represents the electrical activity of the heart. If it is disturbed, the heart's pumping action will be disturbed as well. The heart is a large muscle, and the electrical phenomena of depolarization and repolarization that are related to muscular contractions can be measured on the skin above the chest. Figure 1.31 shows one period of a normal ECG. The depolarization of the atria is visible as the P-wave; shortly afterwards, the atria will start to contract and fill the ventricles. The QRS-complex shows the depolarization of the ventricles; shortly afterward, the ventricles will start to contract and force the blood into the arteries. At approximately the same time, the atria repolarize, but the large amplitude of the QRS-complex makes this invisible in the ECG. The T-wave represents the repolarization of the ventricles.

[1] This does not mean that the ventricle ejects *all* of its blood (see Figure 1.30).

[2] Instead of an *elastance*, we could also talk about a (time-varying, active) *compliance*, which would simply be its inverse.

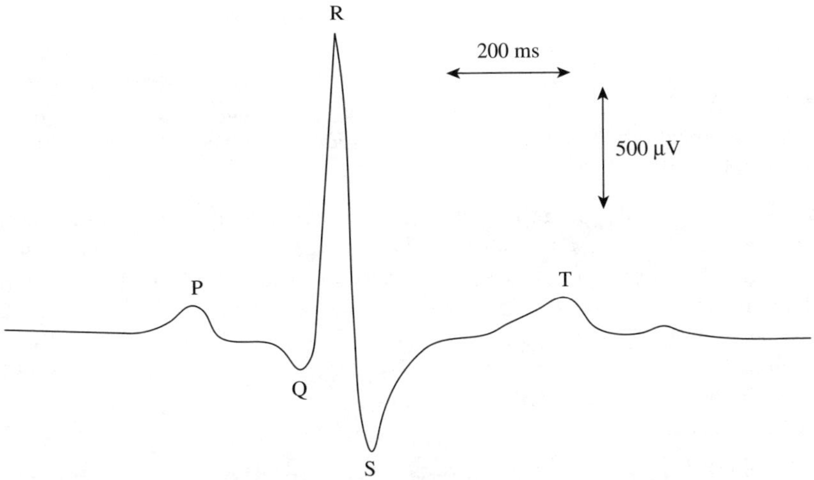

Figure 1.31. One period of a normal electrocardiogram.

If all is normal, each contraction of the heart is visible in both the ECG and the arterial pressure. Sometimes this one-to-one correspondence fails, especially when the heart's rhythm is irregular. This condition is called electromechanical dissociation. If, for instance, a second depolarization of the ventricles so rapidly follows a previous one that the ventricle has not yet been filled with (much) blood, no (noticeable) pressure pulse can be measured although a QRS-complex is visible in the ECG. Such a "premature beat" is not effective and indicates a problem. Therefore, the heart rate determined from the ECG may not be the same as the heart rate determined from the arterial pressure.

Often, special features of the ECG need to be detected. Arrhythmias are rhythm deviations; they do not only show irregular intervals between QRS-periods but also different morphologies (waveforms) of the ECG, especially the QRS-complex. The onset of minor arrhythmias, e.g., premature or ectopic (extra) beats, can signify that ventricular fibrillation, a major arrhythmia, may soon occur. The height (elevation or depression) of the ST-segment (the ECG signal part between S-wave and T-wave) has been recognized as an indicator of cardiac hypoxia (insufficient oxygen supply to the heart).

1.3.7. Some pathophysiologies

Circulatory problems can have two causes: an inadequate pumping function of the heart and/or an inadequate distribution of the blood to the organs.

In *ventricular fibrillation*, the contractions of the individual muscle fibers of the heart are not coordinated into a well-defined depolarization of the heart muscle as a whole. As a result, the heart muscle "flutters" (i.e., small areas contract and relax in an uncoordinated way) and there is no net pumping action. Ventricular fibrillation is fatal if not corrected (see Section 8.4) within a few minutes. It can be caused by a severe electric shock or by irritation of the heart tissue, which can occur in damaged or dying tissues, e.g., due to an infarction. Irritation of the heart tissue can also occur when it is touched by a catheter. Ventricular fibrillation can be detected in the ECG, which loses the characteristic shape of Figure 1.31 and starts to resemble "noise." It can also be detected in the arterial pressure; it loses its pulsations and its mean value drops dramatically.

In *atrial fibrillation*, the pumping action of the atria is lost. Although the ventricles still function normally, their pumping action is less efficient. They are filled less well and therefore pump less blood. This condition is not fatal. In fact, some people may have it for long periods of time without discovering that there is a problem; but they are incapable of strenuous exertion and they get tired easily. Atrial fibrillation can be detected in the ECG, in which the P-wave loses its fixed relation to the QRS-complex. It can also be detected in the arterial pressure, whose pulsations will have smaller and less uniform amplitudes.

Atherosclerosis[1] is a disease of the arteries characterized by fatty deposits (plaques) on their inner wall. This may have two immediate effects: the thickening of the arterial wall leads to a lower compliance, and its decreased diameter leads to a higher resistance. Both will lead to higher blood pressures.

Question: Why does a lower arterial compliance lead to a higher arterial blood pressure? Why does a higher arterial resistance lead to a higher arterial blood pressure?

Since the heart pumps less efficiently at high blood pressures (Figure 1.30), organs are less well perfused. If the atherosclerosis is slight, the patient may have no complaints. If it is severe and present in the blood supply to vital organs, (parts of) these organs will chronically receive insufficient oxygen. As a result, they may deteriorate. In particular, when muscle tissue dies, it is replaced by noncontractile tissue. If this happens in the heart muscle, its contractility suffers.

Problems may arise unexpectedly. Plaques or blood clots may suddenly become loosened from the arterial walls and carried by the bloodstream until they become stuck in small arteries. If they lodge in arteries which supply the heart, myocardial infarction (heart attack) and ventricular fibrillation may be the result; if they lodge in arteries which supply the brain, they may cause a stroke (brain attack). Other organs may, of course, be equally affected, although the symptoms may be less dramatic.

QUESTIONS

1. Describe the major functions of the respiratory and circulatory systems.
2. Describe the major components of the respiratory and circulatory systems and how they interact.

[1] Arteriosclerosis (literally: hardening of the arteries) is a generic term for a number of diseases which are characterized by a thickening of the arterial wall and a decreased elasticity. Atherosclerosis is one of these diseases.

3. What is diffusion and where is it important?
4. Which muscles play a role in respiration and how do they cause gas transport?
5. Why is it important that the inner surfaces of the alveoli are covered by surfactant?
6. What is the difference between the static and the dynamic compliance? How can both be measured?
7. Which factors determine the oxygen supply to the tissues?
8. Describe the oxygen and carbon dioxide partial pressures as these gases pass from outside air to peripheral cells and back.
9. Which factors determine the oxygen and carbon dioxide content of the blood?
10. What is the role of dead space in humans? Research the role of dead space in animals with very long necks (e.g., giraffe, swan).
11. Which role do airway resistance and lung compliance play in lung pathophysiologies?
12. What are the names of the chambers of the heart and the large arteries and veins that connect to them?
13. Which problems arise when the hematocrit has an abnormal value?
14. Describe the maximum and minimum values and the waveforms of the blood pressures throughout the circulation.
15. Describe the volume of and the pressure in the left ventricle as functions of the time.
16. Give a simple model that describes the relationship between left ventricular pressure and aortic pressure.
17. Explain the shape of the left ventricular pressure-volume loop.
18. Explain how the ECG relates to volume changes of the chambers of the heart.
19. What is the effect of ventricular fibrillation on a patient's well-being? And what is the effect of atrial fibrillation?

2 Physics of Fluid Transport in Tubes

In both the respiratory and the circulatory systems, transport — of gas and blood, respectively — is through tubes. We therefore need to study the transport phenomena in tubes more closely. Fluid dynamics is the branch of science which attempts to understand — and from understanding comes the possibility of prediction — fluids and their behavior. From a fluid dynamics point of view, gases and liquids are both fluids; the only difference between gases and liquids is that their properties (density, for example) have different numerical values.

2.1. FLUID DYNAMICS

We first need to introduce some terms. A *flow line* (or streamline) is the path that a hypothetical small particle would follow if one lets it go somewhere in the flow. The *flow velocity* is the velocity that this hypothetical particle would have; the flow velocity can be different from position to position and can change in time as well. Flow is *laminar* when all flow lines are (almost) parallel (Figure 2.1a); the word *lamina* actually means "layer" or "sheet," and thus refers to a surface rather than a line. A *flow profile* is a two- or three-dimensional plot of flow velocities; Figure 2.1b, for example, shows that when a solid object is placed near the fluid flow, it influences the velocity profile in such a way that the fluid velocity decreases near the surface of the object and becomes zero at the surface. Flow is *turbulent* when the flow lines are not (almost) parallel anymore but intermix (Figure 2.1c).

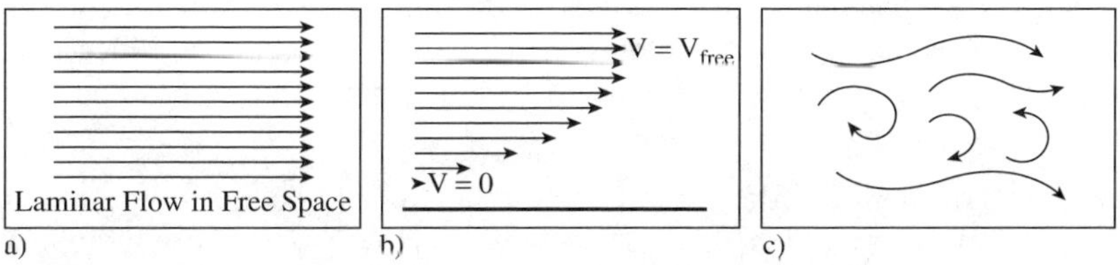

Figure 2.1. Laminar flow in free space has parallel flow lines with equal velocities (a). Velocities are zero at a wall (b). At high flows, flow lines are not parallel anymore but show turbulence (c).

Pressure is created due to the (nearly) elastic collisions of the molecules of the fluid with the surface of a body (e.g., a pressure transducer).

When a fluid is *viscous*, there is *friction* (momentum transfer) between the flow lines or flow layers in the fluid. Fluid viscosity produces *shearing forces* (or simply *shear*). Shear stress is the frictional force between neighboring flow layers or between a flow layer and the vessel wall. The difference between a solid and a fluid is that a fluid does not resist being "sheared": whereas it is hard to compress a solid by a unidirectional force, a fluid just moves away. Fluids do, however, resist variations in the rate of change of shear: they are viscous; that is, one fluid layer tends to take neighboring layers along in its flow. Greater resistance to variation in shear rate of change means higher viscosity: in that case the fluid is more "sirupy."

In a laminar flow, viscosity ensures that neighboring flow lines have near identical velocities (see Figure 2.1b). Sometimes the region of high shear is confined to a thin layer of fluid (the "boundary layer") near the wall of the vessel through which it flows. If that is the case, then outside this layer the fluid behaves as if it were inviscid. Even if the fluid is viscous, the assumption of zero viscosity will be acceptable if the boundary layer is thin compared to the radius of the vessel.

The derivation of the equations of motion for fluid particles originates from a number of conservation laws: mass, momentum (or impulse) and energy in a certain volume are equal to the rate at which they enter that volume plus the rate at which they are created (or destroyed) inside that volume minus the rate at which they leave that volume.

To make this less abstract, we will give an example of a mass conservation law. For the mass of oxygen m_{O2} inside the lung one can write, for an infinitesimally short time period dt

$$dm_{O2}/dt = dm_{O2in}/dt - dm_{O2out}/dt \tag{2.1}$$

where dm_{O2} is the change in O_2 mass during dt, dm_{O2in} is the O_2 mass entering the lung during dt, and dm_{O2out} is the O_2 mass leaving the lung during dt. Since we know that no oxygen is created or destroyed in the lung, terms to describe creation or destruction of O_2 mass are not needed.

The equations of motion for a general fluid are extremely complex and impractical to solve. Therefore, a number of — often very accurate — simplifying approximations are introduced. The first assumption is that the fluid is composed of so many small particles that we can statistically *average the properties* of interest. This works well for gases and liquids under most conditions (but not, e.g., for sand). A second assumption is that the medium can be considered to be *homogeneous*. Another helpful assumption is that the fluid flow is *irrotational*; i.e., there is no rotation of flow lines. Calculations are therefore much easier if we consider flow in straight tubes only. If the flow is *steady* — does not change in time — all time derivatives become zero, again greatly simplifying expressions. We can also assume the flow to be *isentropic* (no entropy change or heat addition) or *adiabatic* (no heat transfer).

We can further often assume that the fluid under consideration is *incompressible*, i.e., has a constant density. Water — and blood — density changes very little with changes in pressure. This is not true for gases. Boyle's "ideal gas" law (see Section 1.2.6) states that

$$P\ V = \text{constant} \tag{2.2}$$

if the temperature is constant. Thus, Boyle's law tells us that a certain mass of gas will be compressed (will occupy a smaller volume) when its pressure rises.

Yet, in many cases we can consider respiratory gas to be incompressible. Differentiation of the above formula gives

$$P\,dV + V\,dP = 0 \tag{2.3}$$

or

$$dV/dP = -\,V/P \tag{2.4}$$

where dV/dP is the compliance of the gas volume under consideration. Inserting a lung volume V of 6 l and an atmospheric pressure P of 1000 cm H_2O, we find — discarding the minus sign — for the compliance of the air in the lung

$$C_{air\text{-}in\text{-}lung} = 6\ l\ /\ 1000\ cm\ H_2O = 0.006\ l/cm\ H_2O. \tag{2.5}$$

When we compare this with the normal compliances of the lung and of the thorax, both of which are around 0.13 l/cm H_2O, we find that C_{air} is only about 5% of their value. Taking the air in the lung as incompressible thus introduces an error in the order of 5% only. In a medical context, such a "small" error is often disregarded.

Depending upon the required analysis, one or more of the above simplifying assumptions may be introduced. When we consider the pulsatile flows that occur in arteries, we may not want to assume that the flow is steady. And when a patient's lung is very compliant, we may not want to consider air incompressible. If, after all allowable simplifications have been introduced, the equations are still too complex to be solved analytically, finite element methods may come to the rescue.

2.2. THE NAVIER-STOKES EQUATION

Normally, some 45% of the blood consists of particles in suspension — red, white and other blood cells. In very narrow capillaries with a diameter about the size of the blood cells, blood flow is very difficult to model accurately. But neither is this necessary: due to their large number, the flow in capillaries — which is mainly determined by the friction of the blood cells with the vessel walls — can successfully be averaged. In larger vessels, the blood flow may be considered *Newtonian.*

A flow is Newtonian when Newton's law applies, which states that the shear stress τ in a viscous fluid is proportional to the gradient of the velocity v perpendicular to the flow direction

$$\tau = \eta\ dv/dr \tag{2.6}$$

where η is the fluid's viscosity (Figure 2.1b shows that the velocity v depends on the position of the flow line).

Now the Navier-Stokes equation applies, which relates pressure and velocity in a constant-entropy, viscous, incompressible Newtonian fluid. In its vector form it is

$$\rho\ \delta\mathbf{v}/\delta t + \rho(\mathbf{v}\cdot\nabla)\mathbf{v} = \rho\mathbf{g} - \nabla p + \eta\Delta\mathbf{v} \tag{2.7}$$

where

ρ = density of the fluid

$\mathbf{v}$ = fluid velocity as a function of x, y, z

∇ = nabla operator (spatial derivative)

$\mathbf{g}$ = gravitation

p = pressure

η = viscosity of the fluid[1]

$\Delta = \nabla^2$

Solutions of the Navier-Stokes equation nicely show the onset of turbulence, the interaction of shear layers, and most other interesting fluid dynamics phenomena. The Navier-Stokes equation, however, still has no general solution, although it can be solved for some practical situations where appropriate boundary conditions exist. It is also a starting point for finite element methods, even though computations may take hundreds of hours on the fastest computers available.

2.3. LAMINAR FLOW AND THE POISEUILLE EQUATION

Since the Navier-Stokes equation is far too complex, we start by introducing a large number of simplifications. Although none of these is true in the respiratory and circulatory systems, they provide a good starting point for understanding. Some of these simplifications introduce only minor differences from what actually happens; other simplifications will be dropped in subsequent sections.

We can derive Poiseuille's formula from the Navier-Stokes equation when we assume a *steady*, *laminar* flow of a *homogeneously viscous* fluid (the viscosity is independent of the shear rate) in a *straight*, *horizontal*, *rigid* tube with a *circular cross section*, where the *friction of the fluid with the wall* is *much larger* than the fluid's internal friction; the latter results in a zero velocity at the wall. Converting to cylinder coordinates — because we have rotational symmetry — we obtain

$$\eta\left[\frac{\partial^2 v}{\partial r^2}+\frac{1}{r}\frac{\partial v}{\partial r}\right]=\frac{\partial p}{\partial x} \tag{2.8}$$

where v is the velocity component in the direction of the tube (all others are zero), r is the radial distance from the center of the tube, and x is the distance along the tube. The boundary condition is zero velocity at the wall: $v(r_0) = 0$. The solution is

[1] The viscosity of blood, measured in blood vessels, may be different from the viscosity that is measured in whole blood. This is due to the inhomogeneity of the blood in the vessel. This inhomogeneity is caused by the finite size of the blood cells. Since there is a velocity gradient across the vessel (the velocity at the wall is zero), the blood cells will tend to roll away from the wall and preferentially move nearer the middle of the blood vessel. The resulting layer of plasma along the wall will cause less friction and a lower apparent viscosity.

$$v(r) = (r_0^2 - r^2)\, P / 4\eta l \tag{2.9}$$

where l is the length of the tube and P the pressure difference over the length of the tube. Note that the flow profile is parabolic.

Integrating over the cross section of the tube, we obtain the flow through the tube

$$F = \int_0^{r_0} v\, 2\pi r\, dr = P\, \pi r_0^4 / 8\eta l \tag{2.10}$$

The above form resembles Ohm's law if we rewrite it as

$$F = P / R, \text{ with } R = 8\eta l/\pi r_0^4 = 8\pi\eta l/A^2 \tag{2.11}$$

where A is the tube's cross-sectional area.

This is Poiseuille's formula. The flow-pressure relationship is linear in l and tells us that, as would be intuitively obvious, the fluid resistance is linearly related to the length of the tube. What may be less intuitively obvious is that the resistance is inversely related to the *square* of the tube's cross section and thus the *fourth power* of its radius. This makes the tube's radius of overwhelming importance for its resistance. Due to the large cross section of blood vessels such as the aorta and vena cava, the resistance of the large arteries and veins is small. Small-diameter vessels, on the other hand, have very high resistances. Poiseuille's formula applies to low flow rates only. At higher flow rates the flow through the tube is no longer laminar but becomes turbulent. At which flow this transition occurs depends on the density and viscosity of the fluid, on the radius of the tube and on the roughness of the inner wall of the tube; making the wall smooth ("streamlining") improves laminarity.

2.4. PULSATILE FLOW AND WOMERSLEY'S SOLUTION

Even if the blood flow in arteries is laminar, it is pulsatile. If fluid velocities vary in time, formula 2.8 complicates to

$$\eta\left[\frac{\partial^2 v}{\partial r^2} + \frac{1}{r}\frac{\partial v}{\partial r}\right] = \frac{\partial p}{\partial x} + \rho\frac{\partial v}{\partial t} \tag{2.12}$$

Moreover, the arterial wall is elastic and pulsates due to the arterial pressure pulse, resulting in more complex boundary conditions. Womersley first solved this problem — which involves Bessel functions — and plotted the resulting velocity profiles. The main finding was that velocity profiles are no longer parabolic but approach flatness ("plug flow") during parts of the heart period, especially in the middle of the tube (Figure 2.2). It was also found that near the wall the flow can even reverse its direction. Similar profiles have actually been measured in vivo. When there is a Womersley flow profile in a tube, it turns out that the flow resistance is again constant, as with Poiseuille flow, but the flow resistance computed according to Womersley's method turns out to be up to a factor 1.5 to 3 higher than the Poiseuille flow resistance. This is an example of how different assumptions (models) lead to different results. It must be realized that both

Poiseuille's and Womersley's results are *idealizations* of a much more complex reality. Which idealization is a better description of reality depends on how realistic the assumptions are in a specific case, and on how accurate we want our model to be. For example, if we know that in practice the flow through a vessel is pulsatile, the assumption that there is "plug flow" at all times often leads to remarkably good approximations, even though this model is extremely simple.

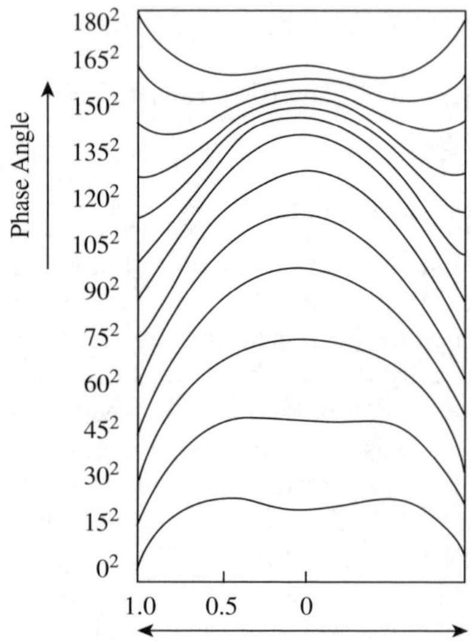

Figure 2.2. In pulsatile flows, the flow profile will at times strongly deviate from parabolic. This figure shows flow profiles when a small constant flow is modulated by a sinusoidal variation. Profiles are drawn at different phases of the sine: the flow profile varies periodically from bottom to top and back again.

2.5. FLUID INERTANCE AND RESISTANCE

When the flow is pulsatile, the mass of fluid inside the tube will have to be accelerated. The acceleration requires a pressure difference Δp per unit length of tube. Consider a tube segment where the flow has a plug profile, i.e., the velocity of the fluid is the same everywhere. The mass inside the tube segment is

$$m = Al\rho \qquad (2.13)$$

and the acceleration is

$$a = dv/dt = d(F/A)/dt = (1/A)\, dF/dt \qquad (2.14)$$

The pressure difference ΔP is then

$$\Delta P = P1 - P2 = \frac{\rho l}{A}\frac{dF}{dt} \qquad (2.15)$$

This formula states that the longitudinal pressure difference P over a length l of a tube is proportional to the rate of change of the flow dF/dt. This proportionality constant is called the fluid inertance L

$$L = \rho l/A \tag{2.16}$$

Fluid inertance is the opposition that the fluid, due to its inertia, offers to acceleration. The formula shows that the wider the tube, the lower the inertance. This may be contrary to intuition; a wider tube contains, after all, more fluid. Yet, this fluid is more easily accelerated, because, in order to reach a certain flow F in a wide tube, its velocity need not become as high as in a narrow tube.

We computed R from Poiseuille's formula as

$$R = 8\eta l/\pi r_0^4 = 8\pi\eta l/A^2 \tag{2.17}$$

Taking the ratio L/R we have

$$\frac{L}{R} = \frac{\rho l}{A}\frac{A^2}{8\pi\eta l} = \frac{\rho A}{8\pi\eta} \tag{2.18}$$

Wider tubes have both less resistance and less inertance than smaller tubes. This formula states in addition, that the wider the tube, the more inertance dominates resistance. In large blood vessels such as the aorta, inertance effects predominate over resistance effects. In small-diameter vessels, on the other hand, resistance effects outweigh inertance effects. The formula also states that the ratio of inertance and resistance depends on the fluid's ρ/η ratio. The value of this ratio for air explains why inertance can be neglected everywhere in the respiratory system.

2.6. THE EULER EQUATION

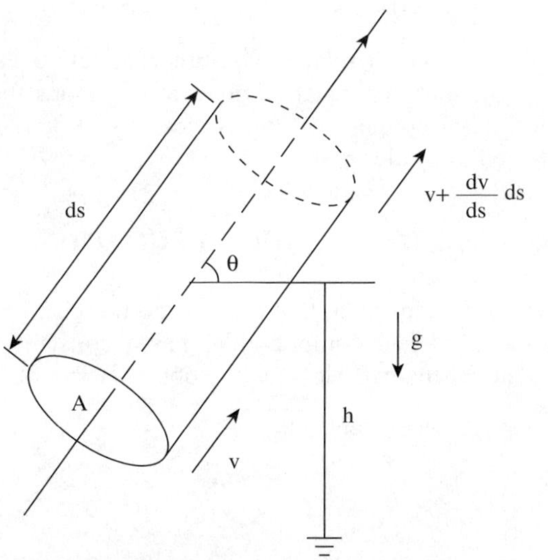

Figure 2.3. Flow through an inclined tube in a gravitational field. For details see text.

Gravity affects flows and pressures. Consider the forces that act on a small volume of fluid that flows in a tube. We assume that the flow is irrotational. The fluid volume has cross-sectional area A, density ρ and length ds, where s is the distance traveled along a flow line. The fluid volume's mass is ρ A ds. It moves along a flow line inclined at an angle θ to the horizontal at a height h in a gravitational field g (Figure 2.3).

Pressure forces act normal to the front and back surfaces of the fluid volume. On the left, the pressure is P, and the accelerating force due to this pressure is P A. On the right the pressure has changed to P + dP = P + (dP/ds) ds and the force to (P + (dP/ds) ds) A. The latter force slows the fluid down. The net force opposing motion is thus –A (dP/ds) ds.

Gravity acts in the vertical direction, but only the gravity component g sin θ opposes motion, resulting in a gravitational force in the direction of s of – ρ A ds g sin θ.

On the left side in Figure 2.3, the fluid's velocity is v. After it has traveled a distance ds, the velocity is v + dv/ds ds. The fluid's velocity change is dv = dv/ds ds, and its acceleration is thus dv/dt = dv/ds ds/dt = v dv/ds.

Now we apply Newton's second law (F = m a) to the fluid volume: its mass times its acceleration is equal to the forces acting on it. There are two forces; one is due to pressure, the other to gravity. Thus

$$\rho\, A\, ds\, v\, dv/ds = -\, A\, dP/ds\, ds - \rho\, A\, ds\, g \sin\theta \qquad (2.19)$$

which simplifies to

$$\rho\, v\, dv = -\, dP - \rho\, g\, dh \qquad (2.20)$$

Rearranging terms we have

$$dP + \rho\, v\, dv + \rho\, g\, dh = 0 \qquad (2.21)$$

This is Euler's equation, which relates pressure changes to velocity changes and changes in height. Note that this equation disregards viscous forces and hence may not be satisfactory for flows where viscous forces cannot be neglected. It also does not include forces due to turbulence.

2.7. THE BERNOULLI EQUATION

If we assume, in the above formula, that ρ and g do not change along the particle's path — i.e., that the fluid is incompressible (has a constant density) — we can integrate the formula over the particle's path to obtain Bernoulli's equation

$$P + \tfrac{1}{2}\,\rho\, v^2 + \rho\, g\, h = \text{constant} = P_{total} \qquad (2.22)$$

The first term stands for the internal pressure[1] in the fluid, which is caused by collisions of the fluid's molecules with the vessel wall. The second term is a "dynamic" (or kinetic) pressure resulting from the fluid's velocity. The third term is the hydrostatic pressure; this term depends upon the height where the measurement is performed relative to the reference pressure.[2] The constant represents the total or "stagnation" pressure that would be measured at zero height ($h = 0$, the "reference height") in a stagnant fluid ($v = 0$).

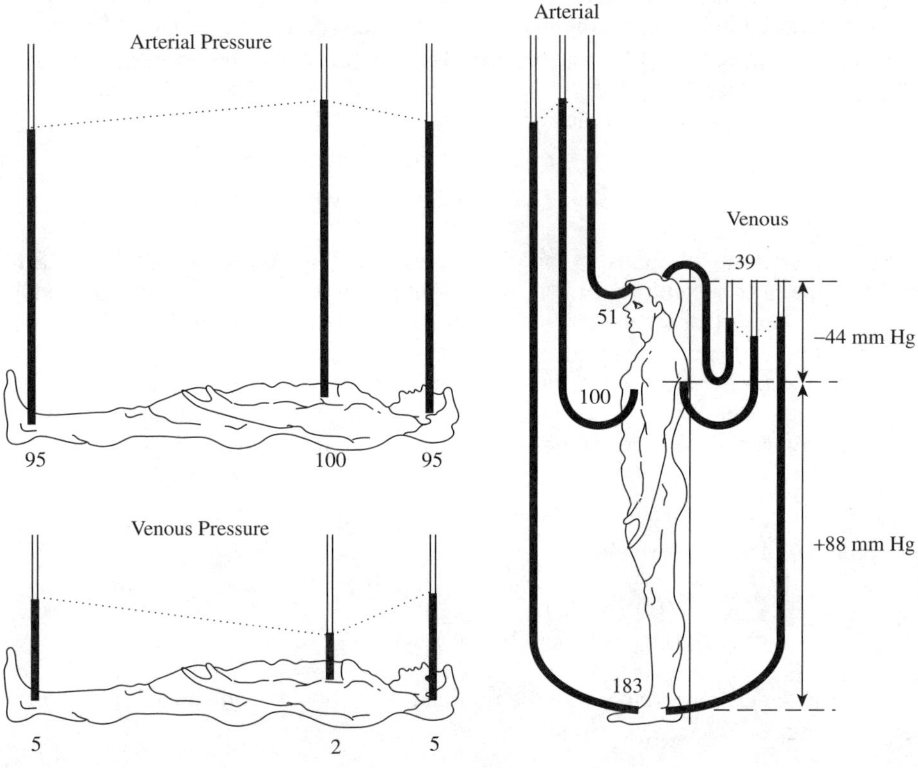

Figure 2.4. Hydrostatic pressure differences are large in an erect person. Zero pressure is conventionally the pressure at the (level of the) heart.

When we talk about respiratory gas pressures and circulatory blood pressures, we implicitly mean the first term, the *internal* pressure. Conceptually, it could be obtained by a very small pressure transducer that is carried along within the flow ($v = 0$ and $h = 0$). In practice, this is usually impossible. The second and third terms, however, can be neglected when $\frac{1}{2}\rho v^2$ and ρgh are small compared to P, or eliminated if they are not.

[1] Since the dimensions of pressure and energy are identical, the terms can also be considered to be different types of energies.
[2] Note that in medicine a pressure measurement is a *differential* measurement, i.e., a difference between *two* absolute pressures. That is why the reference height is important.

In actual gas pressure measurements, the low gas density ensures that the second and third terms can almost always be disregarded. In blood pressure measurements this is not the case. The second term, however, can be eliminated by measuring in such a way that the blood velocity component toward the pressure sensor is zero (see Section 3.2.1.1). The third term cannot be eliminated (we want, after all, to measure pressures at different sites and thus at different heights), but it can be standardized by choosing the reference pressure at the height of the heart (to be more precise: at the height of the tricuspid valve between right atrium and right ventricle). Figure 2.4 shows the large influence of the hydrostatic pressure in an erect person: the arterial pressure measured in the foot is 183 mm Hg. Correction for the term ρgh (88 mm Hg) yields the "true" arterial pressure of 95 mm Hg, only 5 mm Hg less than the arterial pressure measured in the aorta at the height of the heart.

2.8. TURBULENT FLOW

Poiseuille's formula applies to low flow rates only. At higher flow rates the flow through the tube becomes turbulent. But turbulence is not an on-or-off phenomenon. Figure 2.5 shows that when the flow rate is increased beyond a certain critical point, vortices (eddies) appear where the vessel wall is rough or where the flow meets an obstacle. At higher flow rates, the vortices start to detach from the wall and are carried along with the flow, but they still dissipate after a short distance. At even higher flow rates, vortices appear everywhere in the flow. This is called full turbulence.

The (dimensionless) *Reynolds number* Re is an indicator of when flow F starts to become turbulent. Its formula is

$$\mathrm{Re} = \frac{2\rho F}{\pi\eta r_0} = \frac{2\rho F}{\pi\eta r_0}\frac{r_0}{r_0} = \frac{F}{\pi r_0^2} 2r_0 \frac{\rho}{\eta} = \frac{vD}{\nu} \qquad (2.23)$$

where v is the average velocity of the fluid and D is the tube's diameter. The term ν, equal to η/ρ, is called the kinematic viscosity.

Rearranging terms, we find that in a specific fluid (characterized by its viscosity η and its density ρ) in a specific vessel (characterized by its diameter D), turbulence starts at a certain velocity called the "critical velocity"

$$v_{crit} = \frac{\mathrm{Re}\,\eta}{\rho D} = \mathrm{Re}\,\nu / D \qquad (2.24)$$

In highly polished tubes the transition from laminar to turbulent flow can occur at Reynolds numbers as high as 10,000. Under most "engineering" conditions it is somewhere between 2000 and 4000. For the blood flow in arteries to become turbulent, the literature usually gives a Reynolds number of around 2000, but sometimes turbulence in arteries starts at Reynolds numbers as low as 800 to 1200. Before the flow becomes fully turbulent, there is an intermediate Reynolds number region (for water flowing through a long tube: 2100 to 40,000) where the flow is partly turbulent.

When the velocity is increased above the critical velocity, the volume flow — which, when the flow was laminar, was proportional to the pressure difference — increases less rapidly with increasing pressure. It soon becomes independent of the viscosity of the fluid, being dependent mainly on its density. When fully turbulent, the flow is nearly proportional to the square root of the pressure over the tube

$$P \propto \rho v^2 = \rho F^2 / D^2 \tag{2.25}$$

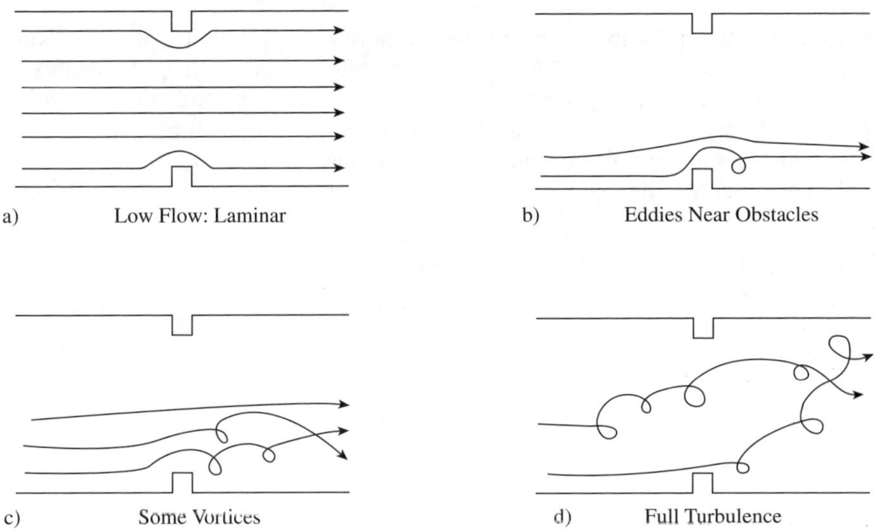

Figure 2.5. At low flow rates, the flow is laminar at an obstacle (a). At higher flow rates, eddies appear (b). At even higher flow rates, eddies become stronger and enter the flow away from the obstacle (c). Finally, flow is fully turbulent (d).

The pressure is now mainly required to overcome the turbulence and to impart kinetic energy to the fluid. When turbulence has not yet fully developed, the flow-pressure relationship is intermediate between Poiseuille's formula and 2.25

$$P = K_1 F + K_2 F^2 \tag{2.26}$$

Turbulent flow is less efficient than laminar flow and, although it exists there, plays no major role in either the respiratory or the circulatory system. Turbulence does occur in the larger vessels such as trachea, main bronchi and aorta, but the diameters of these vessels are so large that pressure drops are small, even though the flow is turbulent; see, e.g., Figure 1.24. In the circulation, turbulent flow is probably significant only in the initial segment of the aorta (also called the "aortic root" or the "ascending aorta") and around the heart valves. Some turbulence also occurs where vessels branch or join. For instance, the soft blowing sound that one can hear in a stethoscope during quiet breathing in a healthy person is said to be related to turbulence as air flows from small air passages into the alveoli.

A method to prevent the flow through a tube from becoming turbulent is to divide the total flow up into a number of parallel smaller flows (see Section 4.1.1 for a measurement device that uses this principle). When, for instance, four tubes each having diameter ½D are used instead of one tube having diameter D, the Reynolds number for each individual tube — and thus also for the four-tube assembly — decreases by a factor 2. Since for the most part the flow in both the respiratory and the circulatory systems is carried by large numbers of small parallel tubes, turbulence can be mostly disregarded.

2.9. FLOW THROUGH COLLAPSIBLE TUBES

Let us consider what happens when the pressure *outside* a tube is higher than the pressure *in* the tube. Figure 2.6 shows a compliant, thin-walled, collapsible tube where a flow F is caused by a pressure drop from P_1 to P_2. Outside the tube we have a pressure P_B, with $P_1 > P_B > P_2$. Inside the tube we have a pressure gradient from P_1 to P_2, and from some point on the outside pressure P_B is higher than the pressure inside the tube. To the right of this point the tube collapses. Several things can happen.

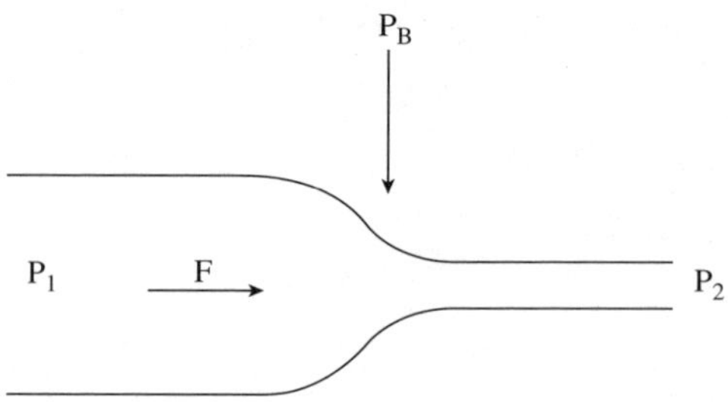

Figure 2.6. A compliant tube is compressed (collapses) where the outside pressure is larger than the pressure somewhere inside the tube. The value of the outside pressure does not matter, as long as it is higher than P_2.

First of all, we note that — to a first approximation — the pressure inside the tube cannot become smaller than the outside pressure, because the external pressure is then simply transferred passively through the tube wall. This means that everywhere to the right of the point where the collapse starts, the pressure inside the tube will be P_B. But this has no consequences for the flow, which is zero because the tube's diameter has become zero. This in turn means that the flow in the uncollapsed section of the tube is zero as well. Due to the zero flow here, the pressure is P_1 everywhere to the left of the point of collapse. This analysis shows that in thin, collapsible tubes the flow will be fully blocked whenever the outside pressure at some point along the tube is higher than an inside pressure. The "Starling resistance" of the tube thus varies from essentially infinite when the tube is collapsed to some finite value when the tube is fully open.[1] This occurs in the respiratory system upon forced expiration, which can cause high intrathoracic pressures (see Figure 1.6). These high intrathoracic pressures can also cause collapse in veins such as the vena cava.

[1] Electrical engineers may note that the "Starling resistance" curve resembles that of a Zener diode.

Note that the outside pressure P_B acts as a switch. With $P_B > P_2$, the flow is zero; with $P_B < P_2$, the tube does not collapse and carries the normal flow. If the total flow is divided over a large number of small tubes and if there is an outside pressure *gradient*, the outside pressure will *modulate* the total flow. This is shown in Figure 2.7, where the outside pressure smoothly increases in the vertical direction from P_{B1} to P_{B2}. If $P_{B1} < P_2$ but $P_{B2} > P_2$, flow will occur normally in the upper tubes but not at all in the lower tubes.

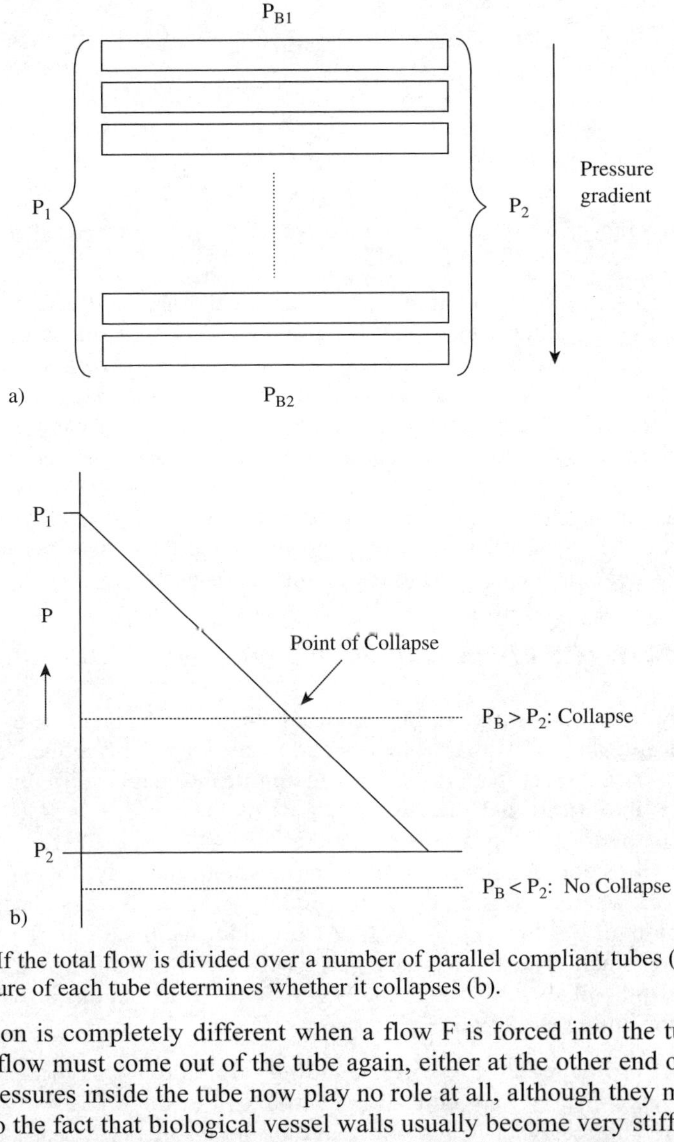

Figure 2.7. If the total flow is divided over a number of parallel compliant tubes (a), the transmural pressure of each tube determines whether it collapses (b).

The situation is completely different when a flow F is forced into the tube (Figure 2.8). This flow must come out of the tube again, either at the other end or through a rupture. Pressures inside the tube now play no role at all, although they may become high due to the fact that biological vessel walls usually become very stiff when fully stretched. This can occur when a patient is ventilated by a ventilator, which produces a constant inspiratory flow. In some cases the resulting pressure may become so high that tissue damage (barotrauma) or even ruptures may occur. In patients with severe atherosclerosis, high pressures and damage can also arise in the initial segment of an extremely incompliant aorta.

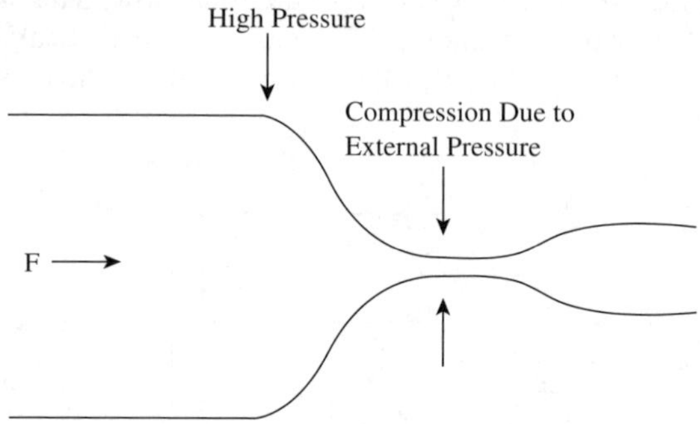

Figure 2.8. If a flow is forced through a constricted tube, high pressures will arise before the constriction.

The flow resistance of a collapsed tube increases dramatically compared to its uncollapsed Poiseuille resistance. Yet collapse plays no major role in the respiratory and circulatory systems, except in isolated cases. There are several reasons for this. First, even thin-walled vessels are compliant (resist volume changes); full collapse hardly ever occurs. In the respiratory system, the larger air ducts cannot collapse due to their anatomical structure. In the circulatory system, the larger veins do collapse, but due to their one-way valves and the fact that they are constantly being filled from the peripheral circulation, this collapse is an asset, not a liability: it transports the blood toward the heart. Moreover, even in collapse the flow resistance of the large veins is negligible compared to the peripheral resistance.

2.10. THE EFFECT OF GRAVITY ON GAS EXCHANGE

Both gravity and collapse play an important role in gas exchange. Due to gravity, the size of the alveoli is not identical throughout the lung. This effect is especially pronounced in an erect patient, where a significant height difference (30 cm) between the apex (top) and the base (bottom) of the lungs exists. This creates a difference in hydrostatic pressure along the lung (Figure 2.9). Since there is a negligible hydrostatic pressure difference in the gas in the lung, the gas pressure is the same everywhere. The pressure in the liquid, however, increases with depth. In the upper parts of the lung $P_{alv} > P_{blood}$. As a result, the alveoli are large and their compliance is small. The alveolar size decreases until the depth where $P_{alv} = P_{blood}$; below this depth, all alveoli are essentially collapsed. Note that this analysis neglects the role of the intrapleural space.

In the model of Figure 2.9, P_{alv} will rise during inspiration. This means that during inspiration the collapsed and more compliant alveoli in the lower parts of the lung expand more and receive a larger percentage of the inspired gas than the alveoli at the apex, whose size cannot increase much due to their low compliance. Thus, the lower parts of the lung contribute most to ventilation. Some of the uppermost parts may not contribute at all; their volume is called the "closing volume."

When the patient lies on his or her back, a similar reasoning applies, but the effect is less pronounced. A similar reasoning applies to the gas flows in the lung's small air ducts (Figure 2.10a). On inspiration, P_{alv} increases with respect to the pressures in the surrounding blood. The air ducts at the lung's apex open first, those at the base last. As a result, the lung is filled from top to bottom. On expiration, P_{alv} decreases and the lung empties from bottom to top. Analysis of the expiratory gas will demonstrate the effect of gravity on unequal ventilation (see Section 6.2).

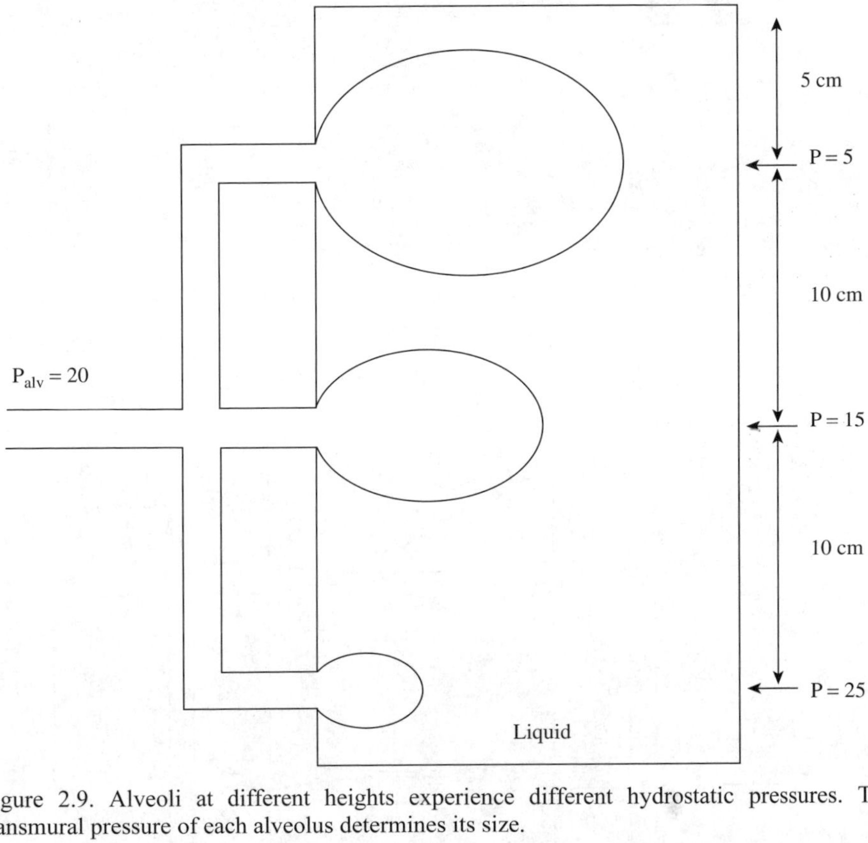

Figure 2.9. Alveoli at different heights experience different hydrostatic pressures. The transmural pressure of each alveolus determines its size.

The lower parts of the lung are also best perfused. This, too, is due to hydrostatic pressure differences, now of the pulmonary artery pressure. Since this pressure is relatively low, a blood column height of 30 cm H_2O in an erect patient is important. The compliant capillaries collapse when their transmural pressure becomes zero. At the top of the lung, the pulmonary artery pressure is lower than the alveolar pressure. In these regions, the arteries are collapsed and conduct no flow (Figure 2.10b).

Further down the lung, the pulmonary artery pressure increases, but the alveolar pressure is still higher than the venous pressure; the capillaries still collapse, although collapse will not be complete.

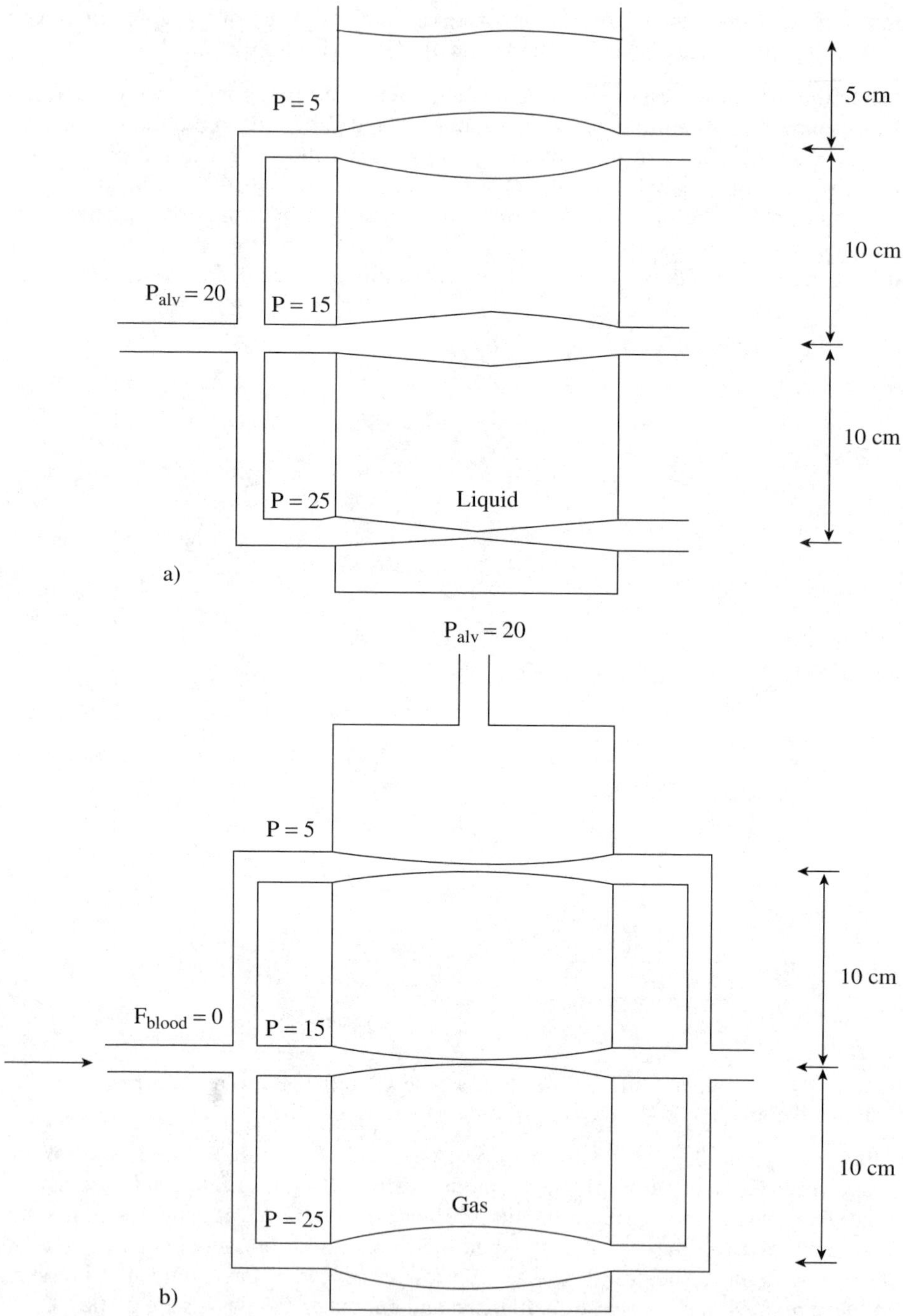

Figure 2.10. A hydrostatic pressure gradient exists in the lung capillaries, but not in the small air ducts. As a result, near the base of the lung air ducts collapse (a), whereas capillaries are more expanded (b).

In the zones at the base of the lung, the venous pressure exceeds the alveolar pressure and no collapse occurs. The overall result is that blood flow decreases almost linearly from base to apex, reaching very low values at the apex. In some diseases, parts of the lung volume may not be perfused at all; these parts are called *alveolar dead space*, and they contribute to physiological dead space.

When the alveolar pressure changes, the depth at which lung capillary collapse starts will change as well. This implies a change of the resistance of the capillary bed (the pulmonary resistance). The pulmonary artery pressure measurement shows indeed pronounced variations due to respiration.

2.11. MODELS

A great number of simplifications have been introduced in our attempts to understand (model) the respiratory and circulatory systems. Given all the approximations, it is actually surprising how well simple models work in many cases. The major reason is that deviations usually are clinically significant only when they are quite pronounced.

Let us review some of our simplifications and why they are allowed:

- The flow is laminar; although this is not true in large vessels and near branches and valves, the overall effects of turbulence are small.
- Tubes are straight; although this is not true anywhere, the overall effect of curvature appears to be small.
- Tubes have constant diameters; although in reality they are often tapered, a constant "effective" diameter per segment is a good approximation.
- Tube walls have a pressure-independent compliance; this is approximately true in the normal operating range.
- Velocity profiles are parabolic; although this may not be true, models remain largely correct although model parameters may have different values.
- Side branches and bifurcations are neglected; these cause difficult to model pressure wave reflections which may result in pressure waveform properties which are quite difficult to explain.

Although reality is so complex that no model, however complex, would be adequate to describe everything in detail, this is neither required nor wanted: complex models usually do not provide better understanding.

Another reason to reject complex models is that, even if an excellent complex model were available, it would probably only model the "average" patient, not the *specific* patient who is being examined. Tuning a complex model to an individual patient would often require more data and longer observation times than are available, if possible at all.

There is a remarkable similarity between fluid transport models and electrical circuits. Fluid transport (flow, pressure) through a section of a tube is governed by the same mathematical equations as electricity transport (current, voltage) through an electrical transmission line. The correspondence is shown in Table 2.1.

Table 2.1. Correspondence between fluid transport and electrical variables.

fluid transport	electricity transport
pressure (difference)	voltage (difference)
volume flow	current
fluid volume in a tube	charge on a capacitor
fluid resistance	electrical resistance
fluid mass	inductance
compliance	capacitance

We recall the following results for a fluid-filled tube of length l: the fluid's inertance $L = \rho l/A$, the fluid's viscous drag resistance $R = 8\eta l/\pi r_0^4 = 8\pi\eta l/A^2$ and the vessel wall's compliance $C = \Delta V/\Delta P$. A long tube can be modeled[1] as in Figure 2.11.

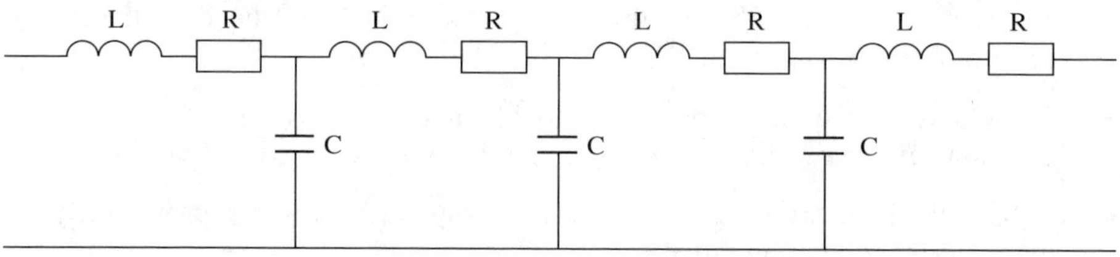

Figure 2.11. Model of a long tube.

A model of a short tube (or a section of a long tube) can take either of the four forms of Figure 2.12, each of which represents a second-order low pass filter. In theory, these models are all equivalent[2]: if many of these sections are connected in series, they all result in the model of Figure 2.11. Non-identical "tubes" can, of course, also be connected in series. This "translation" of fluid dynamics problems into their electrical equivalent allows electrical engineering knowledge and tools to be applied in this area. In Section 3.2.1.1, for instance, we will see how an increase of the arterial pulse pressure in the peripheral arteries can be explained. Some simple

[1] Such a model is identical with the model of an electrical transmission line, where *long* is taken with respect to the wavelengths of the voltages transmitted by the line. We can demonstrate that in the respiratory system, vessels are never "long" in this sense. The circulatory system as a whole is "long" in this sense, but individual vessels (or catheters) are not.

[2] They are *not* equivalent in their input and output impedance. This may provide a criterion for which of the forms to select.

properties of tubes can be recognized immediately: a current injected at the input appears without loss at the output; and the DC voltage at the output of an isolated tube (examined with a high impedance measurement device) is equal to the DC voltage applied to the input.

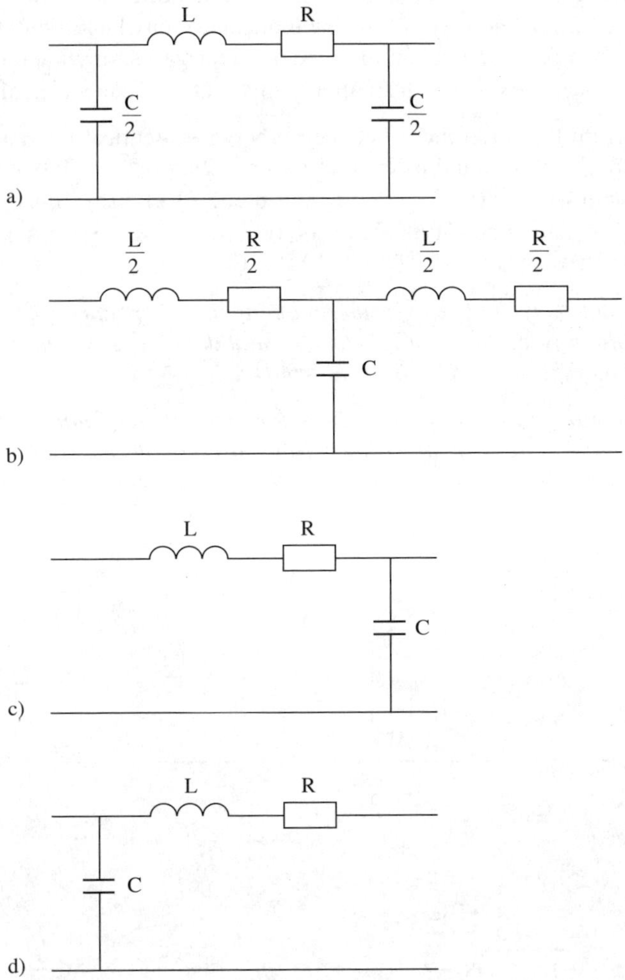

Figure 2.12. A short tube or one segment of a long tube can be modeled in four different ways (a-d).

Question: What do these electrical properties mean in terms of pressure and flow properties?

More simplifications are often possible. They depend on the frequency range of interest, which differs for respiration and circulation. A patient's resting respiration rate is seldom higher than 20 breaths per minute or about 0.3 per second, and heart rate is seldom higher than 200 beats per minute or about 3 per second. A faithful representation of the waveforms encountered requires up to about 10 or 20 (in critical applications up to 50) higher harmonics. Thus, the frequency range of interest in respiration is usually taken to be from 0 to at most 6 Hz, in circulation from 0 to at most 60 Hz.

If the impedance of the vessel wall compliance $1/\omega C$ is small compared to R and ωL, the capacitance can be left out; this can often be done for airflow through tubes and hoses. If the fluid's density is low, the inductance may be left out, because ωL is much smaller than R; this, too, can often be done for gases. In wide blood vessels, R may be left out because it is much smaller than ωL at all frequencies of interest.

Let us investigate the differences between otherwise identical air- and blood-filled tubes more closely. We need the densities of air (1.2 kg/m^3) and blood (1060 kg/m^3), and the viscosities of air (0.019 cPoise) and blood (3 cPoise). Since $L = \rho l/A$, the difference in L is due to density, a factor of 1000! Since $R = 8\pi\eta l/A^2$, the difference in R is due to viscosity, a factor of "only" 150.

Question: Calculate typical (maximum and minimum) impedances R, ωL and $1/\omega C$ for air- and blood-filled tubes of length 10 cm and radius 2.5 mm, using the above data. Assume that the tube's C = 0.01 ml/mm Hg.

Question: Using the results obtained above for both air- and blood-filled tubes, and variant c of Figure 2.12, draw the filter's transfer function (the ratio between output pressure and input pressure as a function of the frequency) in the frequency range of interest.

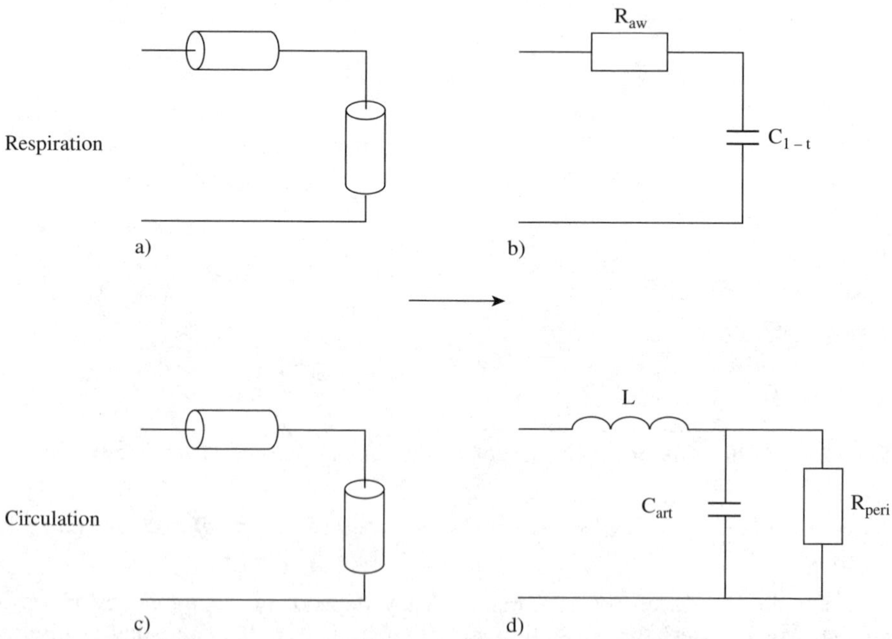

Figure 2.13. Two "tubes" model both respiration and circulation. The simplest model of the respiratory system consists of one R (airway resistance) and one C (lung-thorax compliance) only. The simplest model of the systemic circulation consists of an L (blood inertance), an R (blood flow resistance of the periphery) and a C (compliance of the great arteries).

There is a general technique to build respiratory and circulatory models using electronic networks. First, each "short" tube and each segment of a "long" tube is represented by one of the filters of Figure 2.12. Next, analysis shows which filter components can be left out; this simplifies the filters, where possible. Now, by connecting filter inputs and outputs to each other in the same configuration as a respiratory or circulatory network of tubes, we can model any configuration of them. Simple models need only a few sections. Complex, more accurate models generally consist of many sections. *How complex the model must be depends on the type of questions it should answer.* Too simple models might be inadequate; a comparison of model simulations with measurements in patients must establish their usefulness. This is called *model validation.*

Figure 2.13 shows some extremely simple models. The complete respiratory system is reduced to two "tubes," the *airways* of which only the resistance is represented, and the *lung* of which only the compliance is represented. This results in the familiar model of Figure 1.9. Similarly, the systemic circulation from aorta to vena cava is reduced to two "tubes," the *large and medium-sized arteries* of which the L and C are important, and the *small peripheral vessels* of which only the R is important. This resembles the familiar model of Figure 1.26.

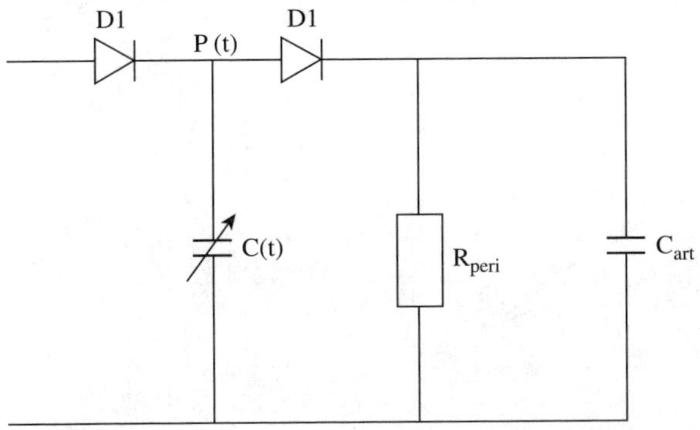

Figure 2.14. A model of the systemic circulation that includes the left ventricle and the heart valves but disregards blood inertance.

Some non-linearities can be modeled relatively easily. We saw in Section 1.3.5 that ventricular pressure P (t) is roughly proportional to ventricular volume V (t) through a time-varying elastance E (t)

$$P(t) = E(t)\,(V(t) - V_0) \tag{2.27}$$

We can rewrite this, replacing 1/E (t) by C (t), into

$$V(t) = V_0 + P(t)\,C(t) \tag{2.28}$$

When we model the valves of the heart as diodes, we obtain the model of Figure 2.14. The left ventricle's contraction is represented by a variable capacitance C (t) and the systemic circulation by an arterial compliance and a peripheral resistance. Analysis is simple as long as the volume does not change. Initially, D1 conducts but D2 does not. When the value of C starts to decrease, P must rise. The first effect is that diode D1 will stop conducting. At some value of C (t), P (t) will become so high that diode D2 starts to conduct.

Question: Repeat this analysis in terms of what happens in the heart.

2.11.1. Natural frequency and damping ratio

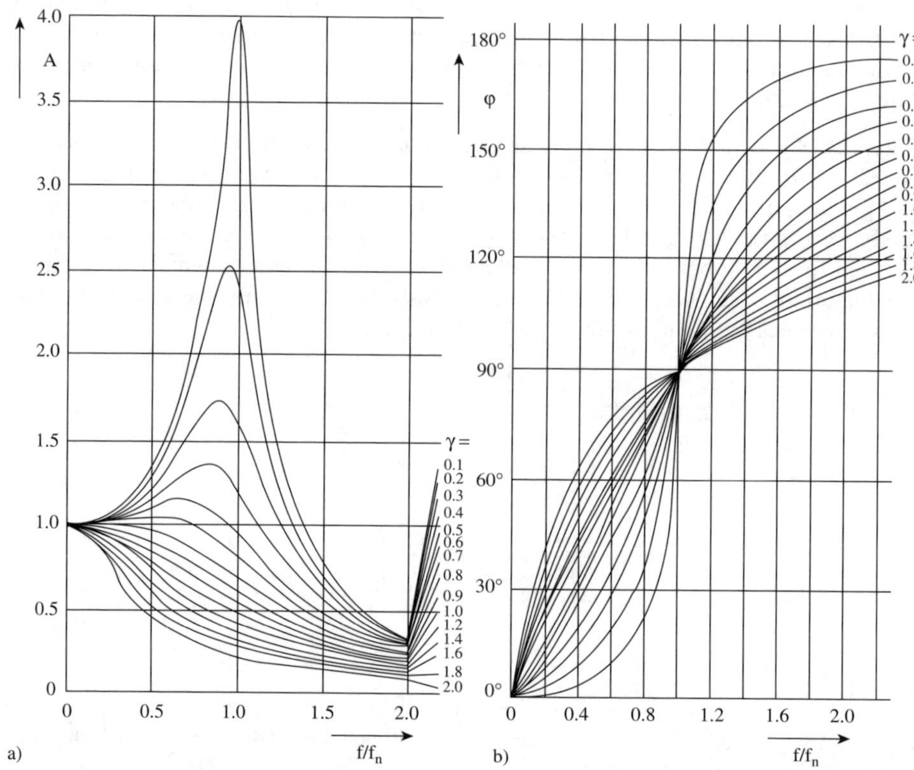

Figure 2.15. Amplitude A (a) and phase φ (b) characteristic of a second-order system.

We will analyze the filter of Figure 2.12c in more detail. As with any second-order filter, the ratio of the voltage (pressure) variations at the output to the voltage (pressure) variations at the input can be fully characterized by only two parameters, the natural frequency and the damping ratio. The *natural frequency* f_n (the resonance frequency of the undamped system) is given by

$$f_n = \frac{1}{2\pi\sqrt{LC}} \tag{2.29}$$

and the *damping ratio* γ (the ratio between actual and critical damping) by

$$\gamma = \frac{R}{2}\sqrt{\frac{C}{L}} \tag{2.30}$$

If $\gamma < 1$, as may happen in arteries, the resonance frequency of the damped system can be computed as

$$f_{res} = f_n \sqrt{1-\gamma^2} \tag{2.31}$$

Figure 2.15a shows the filter's *amplitude* transfer function, where the frequency is normalized with respect to the natural frequency. For $\gamma < 0.7$ *amplification* ($A > 1$) occurs in a frequency range around $f = f_n$, and for $f >> f_n$ the transfer goes to zero. The frequency at which the largest amplification occurs is given by

$$f_{max} = f_n \sqrt{1-2\gamma^2} \tag{2.32}$$

and this maximum amplification itself is

$$A_{max} = \frac{1}{2\sqrt{\gamma^2-\gamma^4}} \tag{2.33}$$

If the amplification is not the same for all frequencies that are present in the input signal, there is distortion of the signal. There is never amplification for $f << f_n$; distortion starts at frequencies around f_n. In man-made systems, one therefore tries to choose $\gamma \approx 0.7$ (this ensures a maximally flat bandwidth) and f_n about equal to or somewhat larger than the highest frequency that occurs in the input signal. In arteries, however, we do not have a choice and significant amplification occurs in peripherally measured blood pressures compared to the pressure measured in the aorta.

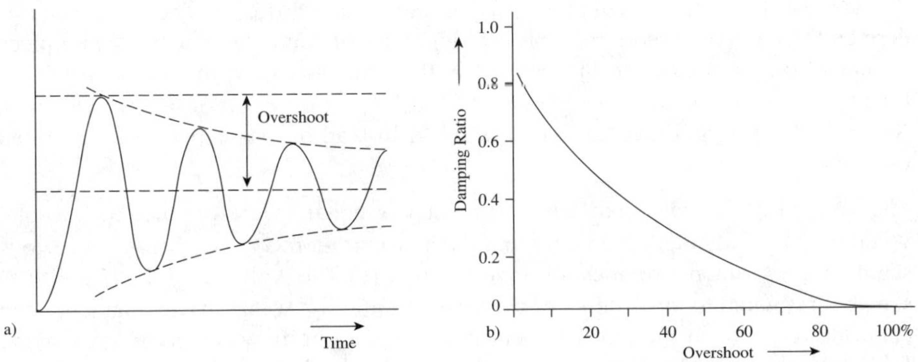

Figure 2.16. Step response of an underdamped second-order system (a) and overshoot as a function of damping ratio (b).

Figure 2.15b shows the filter's *phase* transfer function. Straight phase lines do not introduce distortion of the waveform; they correspond with time delays. When phase lines are not straight, different components of the input signal will have different delays, which may — even with a perfectly flat amplitude transfer — introduce severe distortions. Here, too, maximally straight lines are obtained with $\gamma \approx 0.7$.

Figure 2.16a gives the general form of the filter's *step response*[1] for $\gamma < 0.7$. The response settles in an oscillatory fashion, whose frequency is f_{res}. The *overshoot* (the maximum response value beyond 1) is determined by γ only (Figure 2.16b). At $\gamma = 0.7$ we have an overshoot of only a few percent.

2.11.2. Transport models

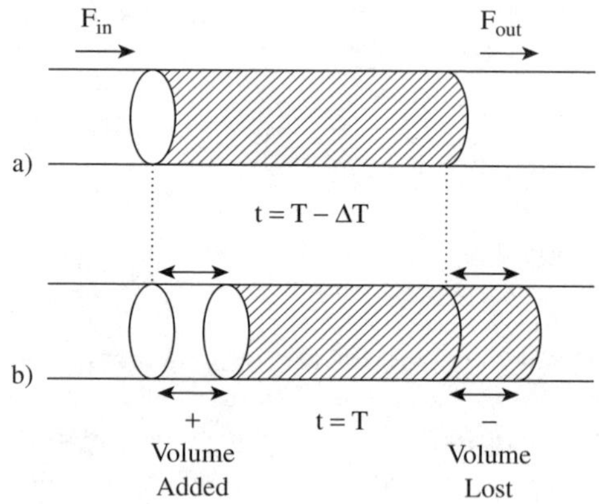

Figure 2.17. In a tube segment, a flow F during a time ΔT adds a volume $F\,\Delta T$ from the left and loses the same volume $F\,\Delta T$ to the segment to the right.

The above models were for pressure-flow relationships. Mass (or concentration, partial pressure, or partial volume) transport models are different. The basic model is derived from the mass conservation law: the mass of some component X in a piece of tube at time t is equal to the mass of X that was present in that piece of tube at time $t-\Delta t$ plus how much X entered the tube during time period Δt minus how much X left the tube during that time (Figure 2.17). Instead of masses, *volumes* are often used in calculations.

The volume of X that enters tube segment i is equal to the product of the total volume $F_{in}\,\Delta t$ that enters segment i and the concentration $C_{X,i-1}$ of X in the previous segment $i-1$, which provides the added volume. The volume of X that leaves segment i is equal to the product of the total volume $F_{out}\,\Delta t$ that leaves segment i and concentration $C_{X,i}$. If the tube segment is compliant, the flows entering and leaving the segment may be different.

$$\Delta V_{X,i}(t) = V_{X,i}(t) - V_{X,i}(t-\Delta t) = F_{in}\,\Delta t\,C_{X,i-1}(t-\Delta t) - F_{out}\,\Delta t\,C_{X,i}(t) \qquad (2.34)$$

Mixing is assumed to be instantaneous (within Δt), so the concentration of X in a segment at any time is equal to the volume of X in that segment divided by the total volume of the segment.

[1] A step response is defined as the response to a stepwise input change. That is, the input is suddenly changed from one constant value to another.

$$C_{X,i}(t) = V_{X,i}(t) / V_i(t) \tag{2.35}$$

This leads to

$$\Delta V_{X,i}(t) = F_{in}\,\Delta t\,V_{X,i-1}(t-\Delta t) / V_{i-1}(t-\Delta t) - F_{out}\,\Delta t\,V_{X,i}(t) / V_i(t) \tag{2.36}$$

where again $\Delta V_{X,i}(t) = V_{X,i}(t) - V_{X,i}(t-\Delta t)$.

If the segments have identical volumes and their compliances can be neglected, we can simplify this expression: each segment's volume V is identical and independent of t, and $F_{in} = F_{out} = F$.

$$V_{X,i}(t) = V_{X,i}(t-\Delta t) + F\,\Delta t\,V_{X,i-1}(t-\Delta t) / V - F\,\Delta t\,V_{X,i}(t) / V \tag{2.37}$$

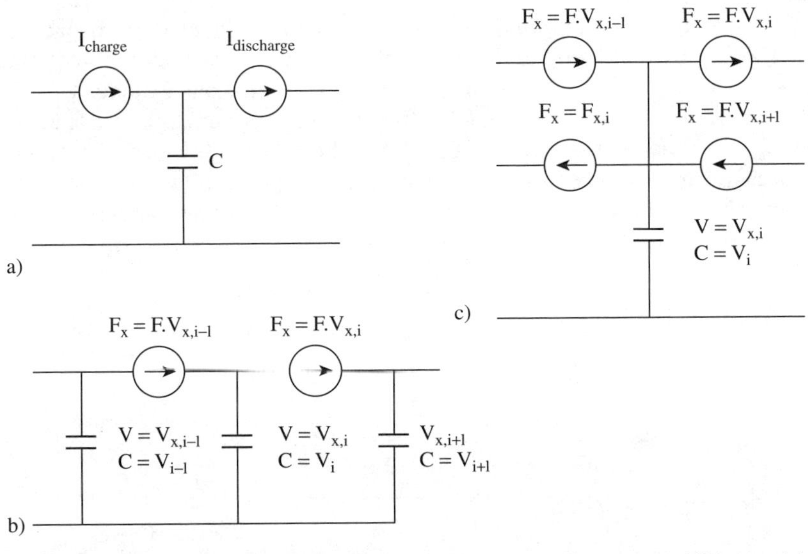

Figure 2.18. A tube segment's volume is modeled by a capacitance C. Addition and loss of volume are modeled by current sources (a). Each current is a product of total flow and concentration (b). Bidirectional flows require extra current sources (c).

We now use the analogy between this formula and an electric circuit in which a capacitor, which has a certain voltage, is charged by one current source and discharged by a second current source. Figure 2.18a gives a model. The formula that applies to the capacitor's voltage is

$$V_c(t) = V_0 + (1/C) \int I_{charge}\,dt - (1/C) \int I_{discharge}\,dt \tag{2.38}$$

If we consider the capacitor's voltage change in a short time interval Δt, during which the currents I_{charge} and $I_{discharge}$ do not change, we obtain

$$V_c(t) = V_c(t-\Delta t) + I_{charge}\,\Delta t / C - I_{discharge}\,\Delta t / C \tag{2.39}$$

We have the following analogies:

$V_{X,i} \sim V_c$

$F\,V_{X,i-1} \sim I_{charge}$

$F\,V_{X,i} \sim I_{discharge}$

$V \sim C$

Thus, we can represent the partial volume of X in a segment as a voltage on a capacitor; this voltage has a value between 0 (no X in this tube segment) and 1 (only X in this tube segment). The capacitor is charged by a current $F\,V_{X,i-1}$ and discharged by a current $F\,V_{X,i}$. Each tube segment is represented by a capacitor, whose value represents the total gas volume of that segment. Current sources transport charge from each capacitor to the next one. The charge transport by each current source depends on two parameters, the voltage of the capacitor whose charge is being transferred, e.g., $V_{X,i}$, and the total flow F. This is shown in Figure 2.18b.

Note that the model of Figure 2.18b assumes unidirectional flow; capacitor C_i is charged with a current equal to the product of the total flow F and the voltage $V_{x,i-1}$ on capacitor C_{i-1}. If the flow reverses, it is not the voltage on capacitor C_{i-1} (i.e., the gas concentration in segment i–1) but the voltage on capacitor C_i (i.e., the gas concentration in segment i) that should decide how large the current is (how much gas is transferred). In order to accommodate bidirectional flows, we therefore need to modify the model to what is shown in Figure 2.18c, where the current is $F\,V_{x,i-1}$ when the flow is from left to right and $F\,V_{x,i}$ when the flow is from right to left.

2.11.3. Multiple models

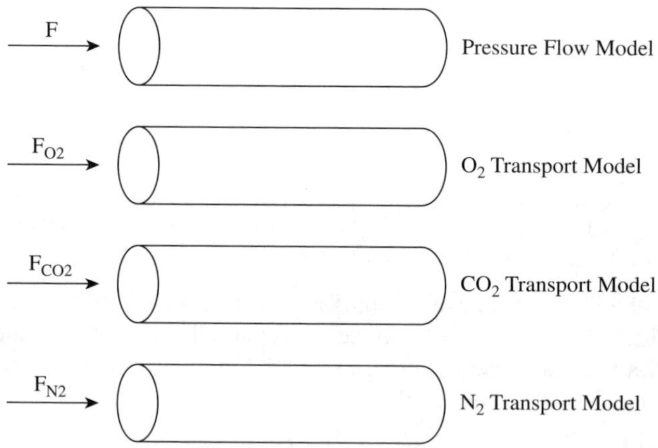

Figure 2.19. A multiple model shows pressure-flow relationships and transport of several gases.

Note that the total flow F in each segment is an unknown in a transport model of a specific gas. A *gas transport model* therefore depends on the availability of a *pressure-flow model* as outlined above. The latter is therefore called the "master" model and the former a "slave" model. It is a "slave" in the sense that it cannot be used by itself but must be used in combination with a model that provides the total flow F for each segment. If, for instance, we want to model the transport of the three important respiratory gases (oxygen, carbon dioxide, nitrogen), we need one "master" model plus three "slave" models. The master models the total flow and the total pressure; the slaves model gas component flows and partial pressures (Figure 2.19).

More complex expressions (and more complex electrical analogies) arise when segment volumes do not have a constant volume. These added complexities may be required to adequately model the lung, the ventilator bellows (see Section 8.2) or the atria and ventricles of the heart. Additions are also required to represent mass *sources* (the lung's production of carbon dioxide; an injection of an indicator into the bloodstream) and *sinks* (e.g., a ventilator's carbon dioxide absorber). Most additions are relatively straightforward and can be modeled as controlled current sources or variable capacitors.

QUESTIONS

1. Explain why it is often allowed in respiratory models to assume that the respiratory gas is incompressible.
2. Which assumptions must hold if we are to assume that the flow through a tube has a parabolic flow profile?
3. Formally derive Poiseuille's formula (formula 2.8) from the Navier-Stokes equation (formula 2.7). Do this by successively introducing simplifications. Make your simplifications explicit in every step.
4. Why is a pulsatile flow often assumed to have a "plug flow" profile?
5. Calculate fluid inertance L and fluid resistance R for a tube of length 1 m and radius 3 cm.
6. How do the values of L and R change when we replace the liquid (water) in a tube with a gas (air)?
7. How do the values of L, R and C of a fluid-filled tube change when a tube's radius doubles? And when its length doubles?
8. What is the hydrostatic pressure difference between head and feet if a patient a) stands up and b) lies down?
9. Why does turbulence hardly play a role in both the respiratory and the circulatory system?
10. Explain how an external pressure can influence the flow through a collapsible tube.
11. Explain how the size of the alveoli depends on their height in the lung.
12. Explain how both the gas flow through small air ducts and the blood flow through small blood vessels in the lung depend on their height.
13. What are the physical determinants of L, R and C when we model flow through a tube as in Figure 2.11?
14. Discuss why the most basic models of the respiratory and the circulatory system (Figure 2.13) are different.
15. Explain why the relationship between ventricular pressure and ventricular volume can be modeled as a time-varying elastance or capacitance.
16. We choose a ventricular model in the form of a time-varying capacitance (Figure 2.14). Derive how the value of this capacitance C (t) varies (approximately) during a full heart cycle.

17. Discuss the general characteristics of a second-order transfer function. What are the natural frequency and the damping ratio? When can amplification occur for certain frequencies? Under which conditions is distortion minimal?
18. Derive difference equation formula 2.37 that describes the transport of fluid through a tube. Derive an equivalent electric circuit that implements the same formula.

3 Measuring Pressure

Pressures are measured with a manometer. Many manometers are essentially a flexible membrane across which a pressure differential is applied. The forces arising from the pressure difference bend the membrane, and sensors that are fixed on or integrated with the membrane (e.g., a strain gauge) convert the membrane's deflection into an electrical signal which is proportional to the pressure difference. Thus, measuring a pressure is simple; getting the pressure to the membrane may not be simple.

3.1. MEASURING GAS PRESSURE

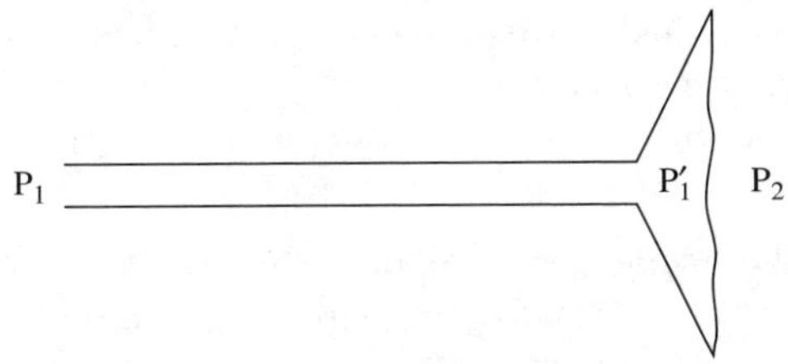

Figure 3.1. The measurement of the deflection of the membrane of the manometer presents the pressure difference $P_1' - P_2$. Pressure P_1' must accurately reflect P_1.

A simple manometer, suitable for gas pressure measurements, is shown in Figure 3.1. The pressure P_1 to be measured is transported from the measurement site to the membrane housing through a tube. The pressure at the membrane is P_1', which is not identical to P_1 due to the fact that a (maybe long) tube is in between. The reference pressure P_2 is atmospheric pressure. So what is actually measured is the difference between P_1' and the atmospheric pressure.

An analysis of the pressure transfer from P_1 to P_1', as has been modeled in Section 2.11, will indicate how faithful the measurement is. In the case of gas pressure measurements, the measurement can be extremely faithful due to a number of factors: the low frequency range of interest makes the effects of the compliances of the measurement tube and of the membrane negligible; the low air density makes inertance effects negligible; and the small excursions of the membrane and hence its low compliance in combination with the low viscous drag resistance of the air in the tube also promise a large bandwidth.

Question: How important is the height at which the transducer is placed and at which the reference pressure is measured?

3.1.1. The esophageal balloon

The mechanical function of the lung depends, as we have seen in Section 1.2.3, on both the compliance of the lung and on the thorax compliance. In order to differentiate between the two, it would be necessary to measure the pressure in the intrapleural space. This, however, is not practically possible. What is possible is to choose an equivalent measurement site just *outside* the lung but *inside* the thorax. At such a site, we would not be able to measure the steady-state pressure in the intrapleural space, but a compliance determination ($C = \Delta V/\Delta P$) requires only pressure changes anyway. As Figure 1.2 indicates, the esophagus provides just such a site. The esophagus is also an extremely compliant tube, so that the pressure outside it will be almost perfectly transferred across its wall. The only remaining problem is how to transfer the pressure inside the esophagus to the tube leading to the manometer.

This is done with the esophageal balloon, which is very flexible yet tends to preserve its own volume. It is sucked empty of air and inserted — usually through the nose — into the esophagus and guided to a position between the two lungs, where it is allowed to resume its volume. Figure 3.2 shows its pressure-volume relationship. Increasing the balloon's volume results in a positive transmural pressure, decreasing the volume in a negative transmural pressure. The balloon's compliance is essentially infinite around the point where the transmural pressure is zero.

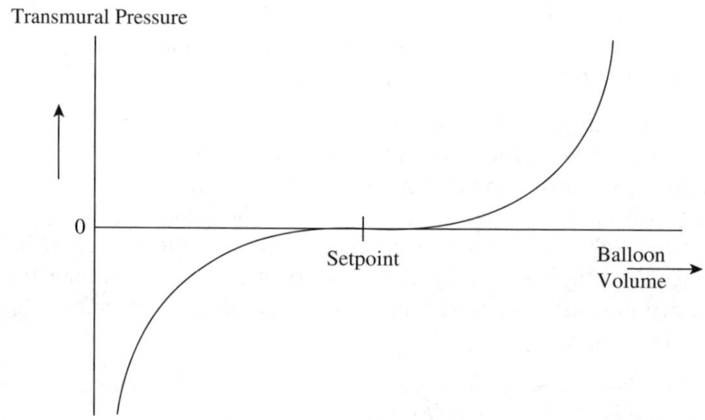

Figure 3.2. Pressure-volume relationship of the esophageal balloon. In its operating point, its compliance is essentially infinite.

Question: Design a procedure that would allow you to determine the lung compliance using the esophageal balloon. Which other measurements would be required? How should the patient be instructed?

3.2. MEASURING BLOOD PRESSURE

Blood pressure is one of the most informative clinically available measures of a patient's condition. Many different measurement techniques exist. The simplest technique is similar to gas pressure measurements (see Section 3.1), but this technique is invasive: it requires access to a measurement site within the patient, where the skin needs to be punctured. Invasive measurements are often unpleasant for the patient; they also present the risks of bleeding, infection and destruction of tissues. Nevertheless, although invasive blood pressure measurements introduce their own problems, they are the most informative measurements available and are therefore indicated in all situations where non-invasive measurements provide insufficient — or insufficiently reliable — information. Criteria for choosing an invasive measurement are the expectation of large, sudden hemodynamic changes, hence the need for a continuous, accurate beat-to-beat measurement, or the need for frequent blood sampling.

3.2.1. Invasive methods

The most frequently used technique to invasively measure blood pressures relies on a hollow needle or cannula which punctures the skin and is brought into the blood vessel whose pressure must be measured. A flexible — but not too flexible — plastic tube is connected to the needle or cannula and transports the pressure to the manometer. In order to prevent blood from flowing into the needle or into the tube where it might clot and introduce a high resistance, the measurement system is perfused by a small flow of a physiological salt solution to which some anti-clotting agent (heparin) is added; this flow is kept so small that it does not disturb the measurement. The catheter is thus filled with what is essentially water, not with blood! Figure 3.3 shows the setup.

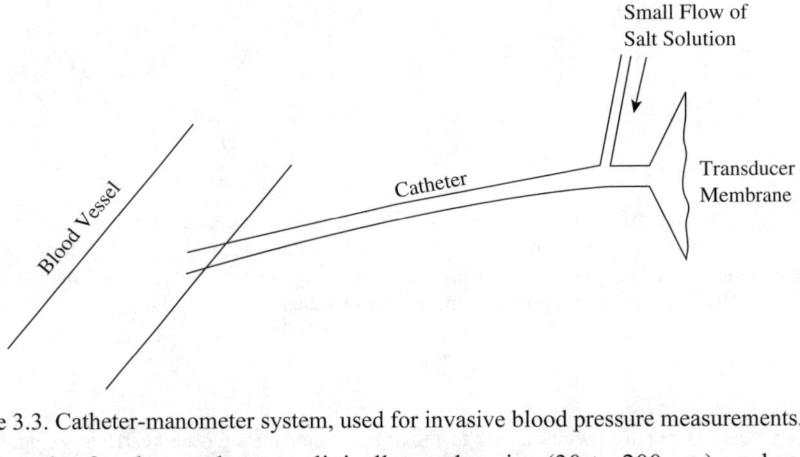

Figure 3.3. Catheter-manometer system, used for invasive blood pressure measurements.

The length of catheters that are clinically used varies (30 to 200 cm), and so does their compliance (stiff to compliant) and their lumen (small to large radius). Long, thin, compliant lines may be easier in use, but short, thick, stiff lines are superior if high quality measurements are required.

Question: Why?

Blood pressures can be measured almost anywhere in the circulatory system, both at the venous and at the arterial sites. Superficial vessels are especially easy to reach. Sites farther away can be reached by pushing a long catheter up or down the vessel until the tip is at the required location. Figure 1.24 illustrates the pressures at the different sites. From a leg artery, for instance, it is possible to reach the aorta and even the left ventricle.[1] This technique does not always work correctly if the vessel branches. The measured blood pressure values and waveforms depend on the measurement site. It is extraordinarily difficult — and considered unethical — to obtain many simultaneous invasive arterial measurements. Figure 3.4 therefore shows the waveforms, obtained at seven different sites, generated by a fairly accurate computer model. Due to the finite pulse wave propagation velocity, more peripheral curves are successively more delayed. Due to the filtering properties of the arteries (which are "natural catheters"; see Section 3.2.1.1), peripheral curves are smoother and have up to 75% higher pulse pressures, equal to that seen in young adults. Due to the smoothing, the incisura or dicrotic notch, which marks the closure of the aortic valve at the end of ejection, disappears.

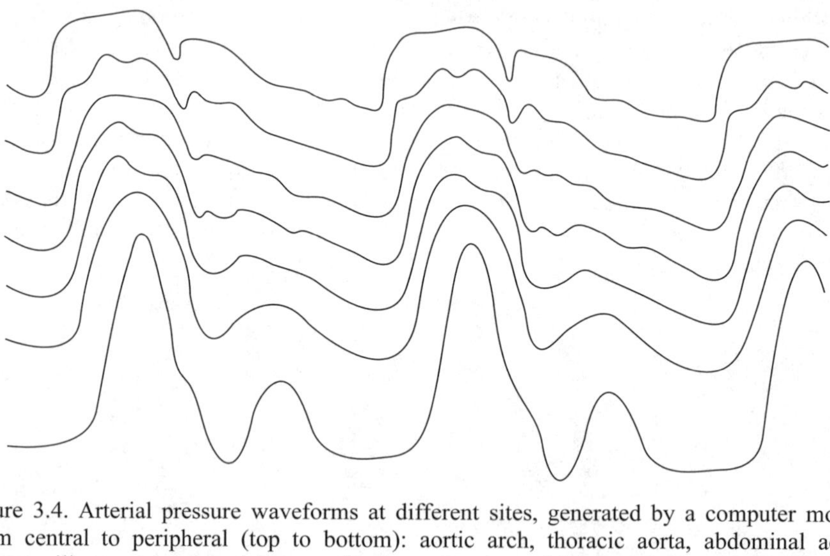

Figure 3.4. Arterial pressure waveforms at different sites, generated by a computer model. From central to peripheral (top to bottom): aortic arch, thoracic aorta, abdominal aorta, common iliac aorta, external iliac aorta, femoral artery, tibial artery. Note the increase in delay and pulse pressure and the loss of higher frequencies.

[1] Extreme care is required when trying to move a catheter through the heart valves against the flow, because this could easily damage the valves. Passing the aortic valve against the flow is, however, a standard procedure.

Question: Although the peripheral systolic pressure is higher than that in the aorta and the diastolic pressure lower, the mean arterial pressure varies little. Explain this.

Another technique is to use a flexible catheter with a small balloon at its tip (a *flotation catheter*, also called a *Swan-Ganz catheter*). This balloon is inflated (with carbon dioxide, which is harmless in case of rupture, because carbon dioxide gas is rapidly absorbed by the blood) and the blood "floats" it downstream, along with the blood. From a vein in the arm, for instance, the vena cava can be reached, then the right atrium, the right ventricle and even the pulmonary artery. Letting the catheter tip move even farther (with a deflated balloon) until it blocks (and then slightly inflating the balloon again), one can even measure the pressure in the lung veins and in the left atrium — these pressures are almost the same — using the blood vessels of the pulmonary circulation as a "natural" extension of the catheter. This is possible because there is no flow (the flow is blocked by the balloon) and because each pulmonary artery has its own vascular bed. Of the left atrial pressure thus measured (the "wedge pressure") only the mean value, however, is to be trusted, since the resistance of the pulmonary vessels "looked through" is high. There is, however, no other way to measure the left atrial pressure.[1]

Question: Qualitatively explain the characteristics of the curves of Figure 3.4 using the theory of Section 2.11.

3.2.1.1. The catheter-manometer system

Catheter-manometer systems have their problems, however. We will analyze the major ones. Since the propagation velocity of a pressure wave through a catheter is around 100 m/s and the maximum frequency component of interest around 30 to 60 Hz, the pressure wavelength in the catheter is at least 2 m. This implies that a one-segment model of a catheter-manometer system is adequate, and that an analysis of the signal transfer from measurement site inside the vessel to manometer membrane has already been fully provided in Section 2.11.1. In contrast to gas pressures, blood pressures often experience considerable distortion on their passage through the catheter: a catheter-manometer system's natural frequency is often significantly lower than 50 Hz, and its damping ratio is also often significantly smaller than 0.7 (see Section 2.11.1). Measured pressures therefore are often an unreliable representation of the true pressures.

Question: Even if the distortions are gross, the mean arterial pressure is measured accurately. Explain this.

One effect of a small damping ratio is that the bandwidth over which the transfer is flat is quite limited. Another effect is that amplification exists, usually for high

[1] Except during diastole, when blood flows from the atrium into the aorta. During this period, the mitral valve is open and the pressure measured in the ventricle is a good reflection of the atrial pressure.

frequencies in the pressure signal (for signal sections where we observe steep slopes). This results in overestimated systolic pressures and underestimated diastolic pressure, and therefore also in inflated pulse pressures. A similar amplification occurs as the blood pressure travels through the arteries from aorta to extremities, because an artery can be regarded as a "natural" catheter. The unreliability of estimating "true" (aortic) pressure variations with a "standard" catheter-manometer system should be evident.

Several solutions to increase the damping ratio to a value of around 0.7 have been described in the literature, none of which is used in clinical practice. These methods are not in use because they introduce their own problems, the most important of which is the extra work and supervision that they require to set up and maintain. Yet, we will describe them briefly. One remedy is called "series damping" and consists of an extra piece of thin (and thus "resistive") tubing (a short capillary) in series with the catheter, either where it enters the blood vessel or where it connects to the manometer. The first position has the advantage that the compensation is catheter length independent; the second is not. The disadvantage of series damping is that it decreases the useful bandwidth of the measurement system. In "parallel damping," an extra resistance (capillary) is connected between a position as close as possible to the manometer membrane and atmospheric pressure. This method does not compromise the bandwidth but introduces an offset error (an error in the DC component of the pressure).

Question: Explain these results using an electrical equivalent of the measurement system, including the extra damping. Alternatively, use a simulation, programmed in, e.g., SPICE.

Another problem of invasive measurements is the presence of small and large but concealed air bubbles in the catheter. Air bubbles introduce an extra compliance. They are a source of distortion if the compliance due to air bubbles becomes comparable to the compliance of catheter and/or transducer. Flushing of the air bubbles should remove them from the catheter (see also Section 9.3). Never flush air bubbles into the patient's circulation!

Question: Why not?

Question: The effect of an air bubble on the transfer characteristics depends on its position along the measurement line. Explain this, and demonstrate where the effect is largest. Do this by modeling the air bubble as an extra compliance and drawing a two-section electrical equivalent of the measurement system, one section to the left of the bubble and another section to its right.

A better method to detect distortion is the "fast flush" technique. A "flush device" can flush a brief, large flow (of fluid, not air!) into the catheter, at the same position where normally only a small flow of a physiological salt solution is flushed into the catheter (Figure 3.3). Normally, the catheter is flushed (into the environment, not into the patient!) when the presence of significant air bubbles is suspected. Their added compliance deteriorates the measurement, and flushing restores the measurement to its original quality.[1] But a flush can be used in a different way as well: a sudden large flush flow creates a large pressure increase at the membrane, which disappears again when the flush ends. The response to this sudden pressure increase and decrease can be recorded; it is the step response of a second-order system (Figure 2.16a). A computer can use the response to estimate the natural frequency and damping ratio of the system (see Section 2.11.1). These two parameters determine the quality of the measurement. An "inverse filter" can then be computed that compensates for the distortion of the catheter-manometer system, if the distortion is not too large. If the computer regularly — or upon inspection of the measured waveform — applies a fast flush, good compensation would be possible regardless of changes in the characteristics of the measurement system that might be due to migrating air bubbles or growing blood clots, provided that the situation does not get too bad. In the latter case an alarm could be issued that a measurement problem exists that must be manually resolved.

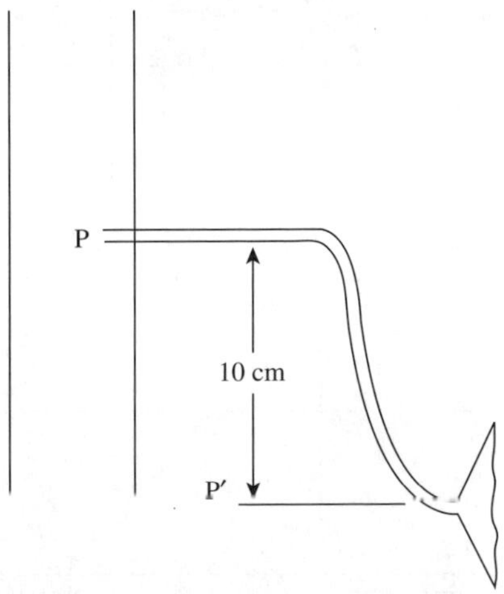

Figure 3.5. If the membrane of the manometer and the measurement site are at different heights, a hydrostatic pressure difference is measured as well.

Bernoulli's equation (Section 2.7) presents us with two further possible error sources. The first is the hydrostatic pressure component that is measured if the catheter is not at the right level. Figure 3.5 shows this. If the transducer is at a vertical level 10 cm below the measurement site, an extra hydrostatic pressure of 10 cm H_2O (the catheter is filled with water) is measured, equivalent to 7.6 mm Hg. Such an error is considerable, even for an arterial pressure.

[1] Flushing needs to be performed whenever a "strange" waveform is measured, since the deviation from normal can be caused by either the patient or a measurement problem. The discrimination between both is important. The frequent flushing that is required often causes clinicians to exclaim that they have to treat the catheter rather than the patient. If a biomedical engineer is present at such an occasion, he or she may be urged to invent better methods which will allow clinicians to return to practicing *patient* care.

The second possible error source is Bernoulli's velocity term. We assume that the catheter's tip has such a small area that it does not noticeably disturb the flow. In Figure 3.6a, the measurement is *against* the flow. The blood flow at the catheter opening is zero, but it is not elsewhere. P is the pressure that we want to measure, but we do measure a *higher* pressure $P' = P + \frac{1}{2}\rho v^2$, where v is the undisturbed blood velocity in the vessel. An example: when the cardiac output is normal, v = 90 cm/s in the pulmonary artery during systole. The term $\frac{1}{2}\rho v^2$ then amounts to 3 mm Hg, whereas P is about 20 mm Hg. The error introduced is thus 15%. With a tripled cardiac output, velocity is also tripled and the term $\frac{1}{2}\rho v^2$ increases ninefold to 27 mm Hg, whereas P has only changed to about 25 mm Hg. The error is now more than 100%!

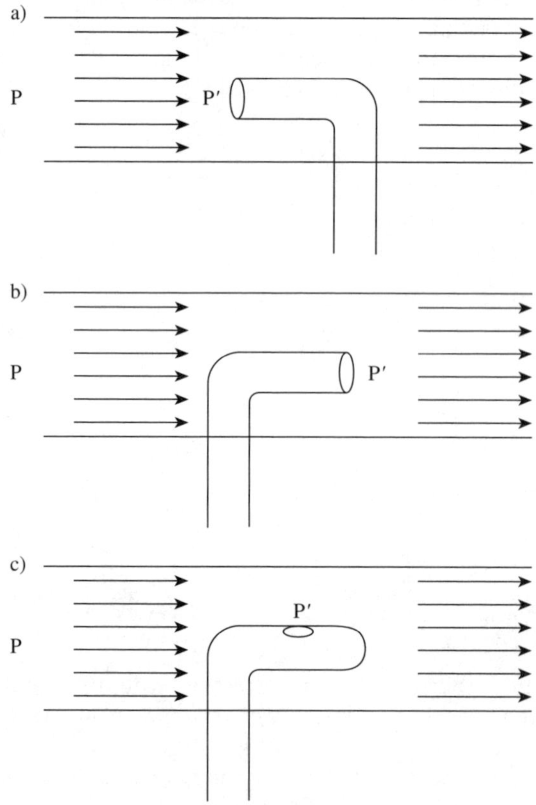

Figure 3.6. Pressure measurement against the flow (a), with the flow (b), and perpendicular to the flow (c).

The situation of Figure 3.6b, where the measurement is *with* the flow, is subtly different. It is unavoidable that turbulence arises at and after the measurement site. Turbulence is characterized by high velocities; the turbulently changing velocities at the catheter opening are most probably higher than the undisturbed velocity v. Bernoulli's equation now tells us that the measured pressure P′ probably will be *lower* than P, although the unpredictability of velocities in a turbulent flow makes it difficult to tell by just how much.

In the situation of Figure 3.6c, the blood velocity *toward* the measurement site is zero. The "side pressure" P′ is now identical with P. This is the correct position of the opening of the catheter. This fact, however, although long known, has not been honored by clinicians and manufacturers yet: catheters with side openings are not in standard clinical use. Nevertheless, at most measurement sites an accuracy of ±5 mm Hg can be expected throughout the measurement range.

Question: Which problem would a side opening introduce?

3.2.1.2. The catheter-tip transducer

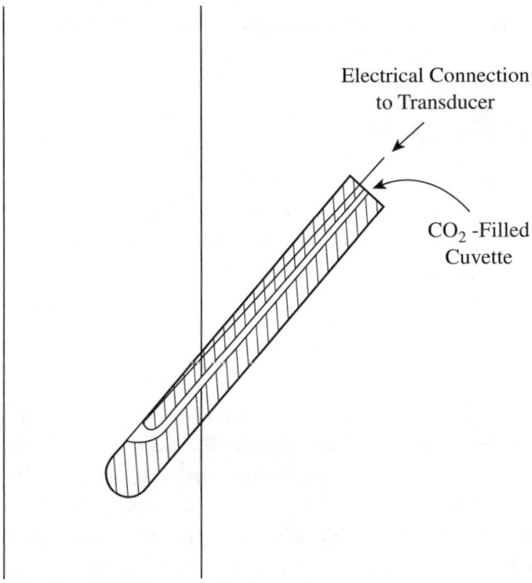

Figure 3.7. A catheter with the pressure transducer at its tip.

Sometimes the bandwidth of a standard catheter-manometer system is not large enough, for instance, in cardiac catheterization measurements, where the heart function needs to be evaluated accurately. We will give an example. The contractility of the heart is defined as the maximum left ventricular dP/dV during the contractile phase of the heart (see Section 1.3.5). Due to the difficulty of measuring ventricular volumes, however, the measurement of the maximum systolic ventricular (or aortic) dP/dt is frequently used as a more convenient indicator of contractility. In order to obtain $(dP/dt)_{max}$ accurately, a high frequency pressure measurement is required. This can be explained as follows. Assume, as a first approximation, that the pressure P can be modeled as $P(t) = A \sin \omega t$. Its derivative is $dP/dt = A \omega \cos \omega t$, and $(dP/dt)_{max} = A \omega$. This value will be obtained only if the measurement system passes angular frequency ω with a gain of 1. If the gain at ω is lower, the measured value of $(dP/dt)_{max}$ will be proportionally lower. P is not a sine wave, of course, but the same analysis applies to each component of a Fourier spectrum.

If a large bandwidth is required, a different type of catheter can be used which has a miniature membrane with embedded transducer at its tip. Figure 3.7 shows a diagram. As in all pressure measurements, the other side of the membrane must be provided with a reference pressure. This catheter, therefore, is hollow as well, but the lumen can now be filled with *gas* (preferably carbon dioxide) instead of a liquid. This eliminates most of the problems that plague liquid-filled systems. Moreover, because only gas has to move back and forth through the catheter's lumen, the bandwidth of the system can be as high as it is in gas pressure measurement systems.

Question: Which problems are eliminated and which are not?

Regrettably, these catheters are very costly and difficult or impossible to sterilize after use. They are therefore only used — e.g., in cardiac catheterization — when it is essential to have a faithful measurement of rapidly changing pressures (steep slopes, large dP/dt) that occur in the heart's ventricles.

3.2.2. Non-invasive methods

A non-invasive blood pressure measurement is to be preferred in routine cases, where the risk of an invasive measurement is not warranted. A correctly executed non-invasive measurement need not be much less accurate than an invasive measurement. Error standard deviations of about 5 mm Hg are mentioned in the literature.

Most methods depend on an inflatable cuff that is wrapped around one of the body's extremities. A high pressure in the cuff is transferred to the underlying tissues and would also tend to cause a negative transmural pressure in the blood vessels below the cuff. But, since these blood vessels are very compliant, they collapse. Blood flow stops completely, indicating that the cuff pressure is higher than the highest pressure that occurs in the blood vessels at any time; this highest pressure is the systolic arterial pressure. On the other hand, if the pressure in the arteries is at all times higher than the cuff pressure, the underlying arteries are never in collapse. The cuff pressure where collapse ends upon deflation is then the diastolic arterial pressure.

The methods differ in how they detect whether blood flows or not. The Korotkoff method detects the sounds that are generated by the turbulence of the flow. Oscillometry detects the small pressure variations that are caused by the opening and closing of the artery under the cuff. And the ultrasound method detects whether particles in the blood (blood cells) are in motion. The accuracies of these methods are comparable.

3.2.2.1. The Riva-Rocci method

The arterial pressure does not vary much throughout the systemic arterial side of the circulation. It is therefore not very important in which artery the pressure is measured. Frequently, the arterial pressure is measured in the upper arm, approximately at heart level. An inflatable cuff, whose internal pressure is measured with some type of manometer (originally a column of mercury, hence the unit mm

Hg) is wrapped around the arm and rapidly inflated above the systolic pressure. The latter is still unknown, of course, but a finger placed on an artery in the wrist or some other method can detect the disappearance of pulsations. Absence of pulsations indicates a cuff pressure higher than systolic. The pressure in the cuff is transmitted to the tissues inside the arm, and thus to the outside of the arterial walls. Whenever the pressure just outside the artery is higher than the pressure inside, the artery collapses and all flow is blocked.

Now the cuff pressure is allowed to slowly decrease. At a certain cuff pressure, pulsations will reappear, because the artery's transmural pressure becomes slightly positive and some flow is passed. But this still happens only when the arterial pressure is close to maximal (systolic). Thus, the cuff pressure, read at the moment of first reappearance of pulsations, equals the systolic pressure.

A finger on a wrist artery is, however, not an accurate enough instrument to detect the first minuscule pulsations; a stethoscope above an artery, e.g., in the elbow, is. Around systole, only a small flow is passed, but the opening through which it passes is also very small and the flow is therefore highly turbulent. This turbulence is audible as *Korotkoff sounds*. The appearance of the first Korotkoff sounds is thus an indicator that the cuff pressure is equal to (better: slightly lower than) the systolic arterial pressure. At lower cuff pressures, blood flows in the artery only if the artery's transmural pressure is positive; the presence of flow is detected as Korotkoff sounds.

When the cuff pressure decreases steadily, blood flows during successively longer periods of time per heart beat. Korotkoff sounds become louder and their character changes. When the cuff pressure becomes lower than diastolic, no turbulence arises anymore and Korotkoff sounds are no longer audible. The disappearance of Korotkoff sounds thus indicates the diastolic arterial pressure. The procedure is sketched in Figure 3.8. Note that the blood volume in the lower arm increases each time the artery passes some blood through. This blood cannot leave the arm because the veins are fully occluded. The pressure in the underarm veins may therefore rise to abnormally high values.

The accuracy of the technique depends on the speed with which the cuff pressure decreases. Since the systolic pressure occurs only about once per second, the cuff pressure decrease per second gives the accuracy: a cuff pressure decrease of 5 mm Hg per second means, that when one systolic value has just been missed, the cuff pressure will be 5 mm Hg below systolic when it meets the next pressure pulse.[1] Yet it is this value that will be recorded as "the" systolic pressure. The accuracy with which the diastolic pressure can be established is the same.

To increase the method's accuracy, it is thus advisable to empty the cuff as slowly as possible. But that would create another problem, as indicated in Figure 3.9. During

[1] This is an idealization. In reality both systolic and diastolic values fluctuate from beat to beat. This, however, does not add to the inaccuracy of the method. It only indicates that the systolic and diastolic pressures delivered by this (or any other) method have a random component.

the procedure, blood flows *into* the arm, but since the cuff pressure is higher than the venous pressure, the veins are continually collapsed and no blood can *leave* the arm.

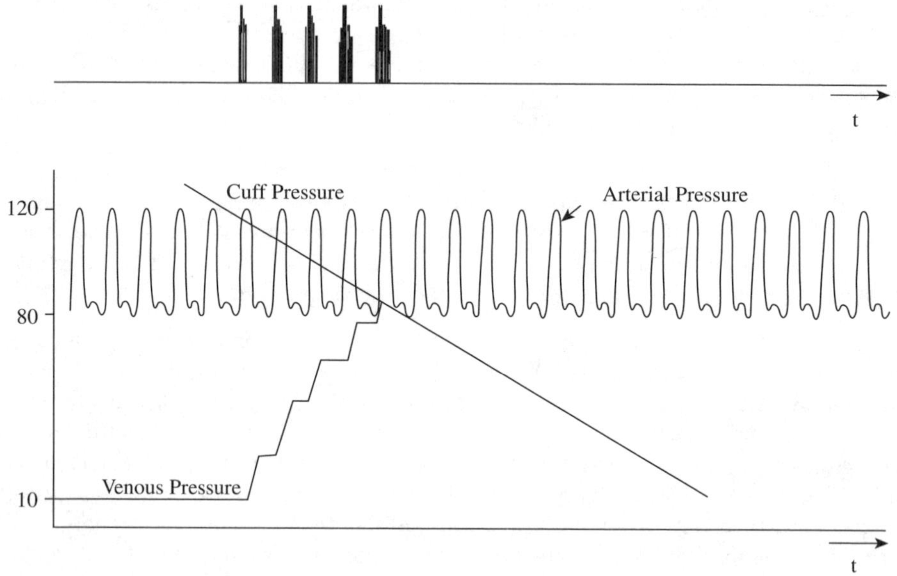

Figure 3.8. Korotkoff sounds appear when some arterial blood flows through the constriction under the cuff. This flow is highly turbulent. Since blood cannot leave the arm, the venous pressure rises. The venous pressure should remain lower than the diastolic arterial pressure.

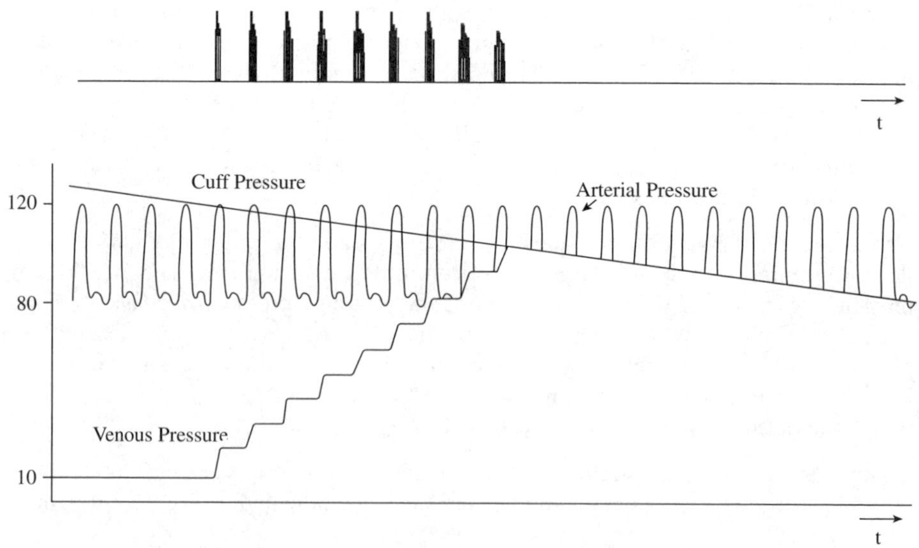

Figure 3.9. Korotkoff sounds disappear when the venous pressure approaches the value of the arterial pressure, resulting in an erroneous reading of the diastolic pressure.

The increase in venous volume causes an increase in venous pressure. Ultimately, the venous pressure in the lower arm would rise so much that it becomes equal to the cuff pressure. At that pressure, venous outflow will start again. The crucial point is that *this pressure may be higher than the diastolic pressure*.

Thus, the venous pressure is now high. The arterial pressure cannot be smaller than the venous pressure. Thus, the pressure difference between arterial and venous pressures is small. Due to this small pressure difference across the artery's occlusion by the cuff, the blood velocity under the cuff is also low. The turbulence of the flow disappears, as well as the Korotkoff sounds. The disappearance of the Korotkoff sounds is normally a sign that the diastolic reading should be taken. Now, however, this is done at a pressure possibly much higher than the original, undisturbed diastolic value. In other words: the obtained "diastolic" value can be completely in error. Thus, although emptying the cuff slowly does increase the accuracy of the determination of the systolic pressure, it makes the estimate of the diastolic pressure unreliable.

Question: The cuff pressure deflation rate is a compromise: neither too fast nor too slow. On which (constant or measurable) factors should this compromise be based?

The accuracy of the auscultatory method is a long-standing concern. If we compile the many references, the systolic pressure may be 1 to, as was found in an extreme case, 112 (mean 6) mm Hg lower and the diastolic pressure 8 to 18 (mean 11) mm Hg higher than the corresponding invasively measured pressures. This is often considered inadequate, for instance, by the Association for the Advancement of Medical Instrumentation (AAMI), whose standard requires an accuracy of ±5 mm Hg for both systolic and diastolic pressure.

With this method, one cannot obtain the value of the *mean* arterial pressure. Although some are of the opinion that the pressure at which the Korotkoff sounds are loudest coincides with the mean pressure, this is incorrect. When nevertheless a mean pressure is needed, a frequently used formula is $P_{mean} = P_{dia} + \alpha\ (P_{sys} - P_{dia})$, where α is usually chosen as 1/3 or 0.4. The accuracy of this heuristically determined formula is doubtful. Theoretically, α should depend on the pressure waveform; for some waveforms, it will be correct, but not for others.

Question: Determine what value α should have if the arterial pressure waveform is modeled as a) a sine; b) a triangle.

The Korotkoff method can be automated for cases where successive measurements are required. In patient monitoring, for instance, one can use a small computer to inflate and deflate the cuff periodically (e.g., every 3 min) and to process the Korotkoff sounds that are now picked up by a microphone instead of a stethoscope.

Several other non-invasive methods are largely identical to what has been described above, but do not rely on Korotkoff sounds but on other indicators of when pulsations start and stop. We will review some of these.

3.2.2.2. The oscillometric technique

The oscillometric method depends on the early finding that the mercury column, which was initially used, starts to slightly oscillate up and down when the arterial volume below the cuff varies. Not only is the cuff pressure transmitted to the artery, but the pulsating arterial volume is also transmitted back to the cuff. The small pressure variations thus introduced in the cuff make the mercury column oscillate. A general finding is that the oscillations start small at the systolic pressure, grow larger until the mean arterial pressure is approximately reached, then decrease again and stop when the cuff pressure reaches the diastolic pressure. Using this method, the systolic pressure is defined as the cuff pressure where the amplitude of the oscillations first reaches a fixed percentage of the maximum, the mean pressure as the cuff pressure where the amplitude of the oscillations is maximal, and the diastolic pressure as the cuff pressure where the amplitude first falls below another fixed percentage of the maximum. Although a physiologic basis for this method has never been established, its errors are comparable to those made by the method based on Korotkoff sounds.

A refinement of the oscillometric method is the sphygmo-oscillographic method. This method uses not only cuff pressure amplitude variations (oscillometric) but also the shapes of the pressure waves (sphygmographic) recorded from a transducer in the cuff.

3.2.2.3. The ultrasound technique

The ultrasound technique depends on the Doppler effect (see Section 4.2.2) to establish whether blood (more accurately: particles in the blood, i.e., blood cells) is in motion below the sensor. When the cuff is fully occluded, no blood is in motion; whenever the intra-arterial pressure is larger than the cuff pressure, blood flow can be detected. The systolic pressure is announced as soon as some motion is detected, the diastolic pressure as soon as motion exists during the full heart cycle.

3.2.2.4. The Peñáz-Wesseling method

None of the non-invasive measurements discussed so far is able to provide the waveform of the arterial pressure. The waveform, however, contains important information, such as the systolic slope (dP/dt), which cannot be derived from non-invasive measurements. Also, the cuff methods give a *single* reading only. Although the measurement may be repeated at regular intervals (e.g., once every 3 min), this may be too infrequent if the patient is very unstable. Also, since the blood flow through the arm is blocked during the measurement, it should not be repeated at a too high rate. For that reason, researchers have long attempted to develop non-invasive methods that can provide the continuous readings of the invasive method. These, too, rely on a cuff, but this cuff does *not* block the blood flow and thus can often be tolerated for longer periods of time.

The body's tissues, except where bones occur or the tissue is too thick, let some light shine through. When a light source is placed at one side of an extremity, e.g., a finger, and a photo detector on the other side, the energy of the light shining through can be detected. If this energy fluctuates, the fluctuations in light absorption can only be due to the varying amount of blood in the tissues between light source and detector, i.e., by arterial pulsations. The pulse oximeter, too, uses this principle.

The method of Peñáz is based on abolishing the fluctuations of the light energy received in the sensor by using a fast feedback system to inflate and deflate a small cuff around an extremity such as a finger (Figure 3.10) in such a way that the light energy received remains constant at some specified level. A constant received light energy implies a constant light absorption by the tissues; this implies a constant finger volume; this implies a constant arterial volume; and this, finally, implies a constant transmural pressure. If the feedback loop is made fast enough, the diameter of the underlying artery or arteries will remain constant at all times. And, more important, the pressure fluctuations in the artery are exactly compensated for by the pressure fluctuations in the cuff. The latter can be measured continuously. Thus, whereas all non-invasive blood pressure measurements thus far have been non-continuous, this method gives a non-invasive continuous blood pressure measurement. The final problem to be solved is that, whereas pressure *fluctuations* are obtained faithfully, the offset (or DC level of the pressure) remains unknown.[1] The artery's transmural pressure is maintained at some constant value, but it should either be maintained at or near zero, or it should be known so that it can be corrected for. Moreover, the maintained transmural pressure can change over time if the patient's characteristics — particularly the tone of the smooth muscles in the arterial wall and the volume of extracellular fluid in the finger — (may) vary rapidly.

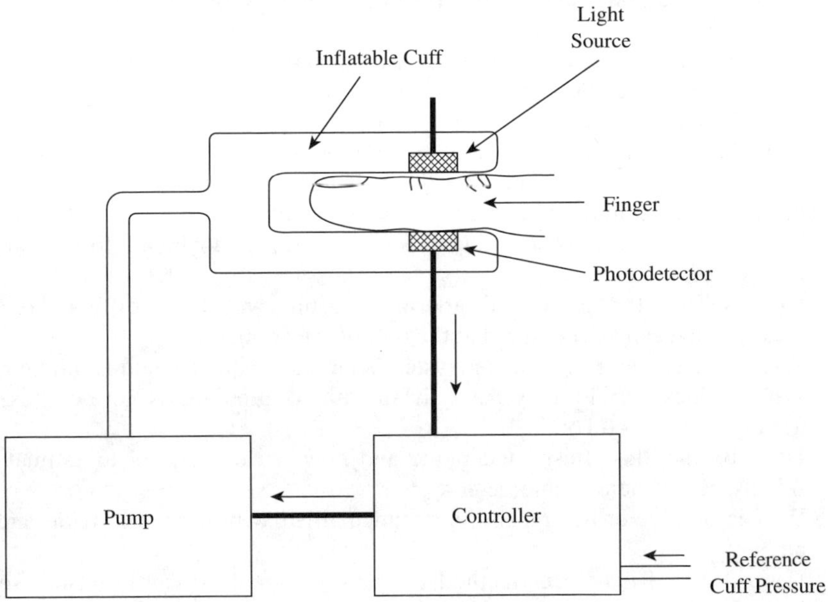

Figure 3.10. A feedback controller varies the finger cuff pressure so that the transmitted light detected by the photodetector has a constant energy. As a result, arterial transmural pressure will be constant.

To solve this problem, Wesseling designed a method that regularly analyzes the waveform of the recorded infrared light output (plethysmogram) in order to find a

[1] This problem also exists when an esophageal balloon is used to measure the intrapleural pressure.

suitable value for the cuff pressure setpoint value. This method reduces the offset error to an average of 1 mmHg, with error extremes of at most 10 mm Hg. This compares favorably with other cuff methods. Theoretically, offset (re)calibrations would need to be performed at a frequency which is high compared to the rate with which muscle tone and fluid volume in the finger vary. In Wesseling's Finapres (FINger Arterial PRESsure) device, (re)calibrations are performed approximately once every minute. During (re)calibrations — which have a duration of two heart beats — the measurement of the arterial pressure is temporarily unavailable.

One problem remains: the pressure is measured at an unfamiliar site, the finger. In cases where extremely high resistances exist in arteries in the arm, the pressure measured in the finger will not be a good reflection of the brachial pressure, which clinicians usually measure. This creates interpretation problems as long as clinicians are not familiar with finger pressures.

QUESTIONS

1. What are the main sources of error when measuring a pressure with a manometer?
2. Discuss the influence of those measurement errors in both gas and liquid pressure measurements if we restrict our interest to low frequencies (lower than 100 Hz).
3. At which anatomical sites can the blood pressure be measured with a catheter-manometer system? Assume that the catheter is long enough. Which sites are difficult or even impossible to reach with a catheter?
4. What is the "wedge pressure," how is it measured, and which information does it provide?
5. Discuss why the peripheral arterial pressure waveform differs from the pressure that can be measured at the root of the aorta.
6. What are a catheter-manometer system's natural frequency and damping ratio? What values should they have if the blood pressure is to be measured accurately up to 40 Hz?
7. Describe the "fast flush" technique and how it can be used to estimate the fidelity of a catheter-manometer system.
8. Why is a catheter-tip transducer's lumen filled with carbon dioxide and not air?
9. Describe the Riva-Rocci method to measure blood pressure non-invasively. Why should the cuff deflation be neither too fast nor too slow?
10. Research the hypotheses of the origin of Korotkoff sounds in more detail, in particular hypotheses about the (near) inaudibility of these sounds at cuff pressure levels where the simple theory given here predicts that one should hear sounds.
11. Describe the Peñáz-Wesseling method.
12. One problem of the Peñáz-Wesseling method is the unknown offset between the average pressure in the arteries and the average cuff pressure. Explain how an occasional Riva-Rocci upper arm measurement can be used to calibrate the Peñáz-Wesseling measurements.

13. Wesseling solves the problem of measuring a finger pressure rather than the brachial pressure that clinicians want by incorporating an electronic filter that computes a brachial pressure from the finger pressure. Discuss which properties this filter should have.

4 Measuring Flow

Whereas measuring pressure (difference) is easy, measuring flow is difficult. A large variety of methods exists, some of which will be described. Several techniques reduce the difficult problem of measuring flow to the easier problem of measuring a pressure difference, using a fixed or variable resistance R and the relationship $\Delta P = R\ F$, which of course holds only for laminar flows. In order to obtain reliable and noise-free measurements, it is important to have a high enough pressure differential. But in order not to significantly disturb the flow, a low resistance must be used. Different devices offer different solutions to the required compromise.

Before the electronic era, it was important for instruments to be linear. If an instrument's readout did not have a linear scale, obtaining the reading could be difficult or lead to errors. Today, this consideration is not important anymore; electronics — or microcomputers — can linearize any function. The importance in sensor research and development nowadays is on reproducibility. Reproducibility of a flow measurement is least guaranteed when flows are turbulent, especially if the turbulence pattern can also be highly variable.

4.1. MEASURING GAS FLOW

To measure gas flow, many methods resort to the much easier problem of measuring a pressure difference across a flow resistance. Some methods ensure a laminar flow; other methods do not.

4.1.1. The pneumotachograph

How can we force a flow to be laminar? An intuitive answer would be to force the flow through a large number of straight small-diameter parallel tubes. This forces the flow "particles" to follow the direction of the tubes, which enforces straight flow lines.

The pneumotachograph (Figure 4.1) follows this basic principle. It measures gas flow by offering a resistance to the flow and measuring the pressure differential across it. Laminarity of the gas flow is ensured by dividing the total flow F up into a large number N of small parallel flows F/N. This decreases the Reynolds number Re of each small flow contribution F/N by a factor of $N^{½}$ to such low values (see Section 2.8) that a laminar flow is guaranteed. Thus, Poiseuille's law (see Section 2.3) guarantees that the resistance R offered to the flow is constant, and that the pressure differential ΔP that is measured is linearly related to the flow F.

According to Poiseuille, the pressure reading is viscosity dependent, and thus also dependent on temperature, pressure and gas composition. Pressure and gas composition variations can normally be disregarded in this method.

Question: Why?

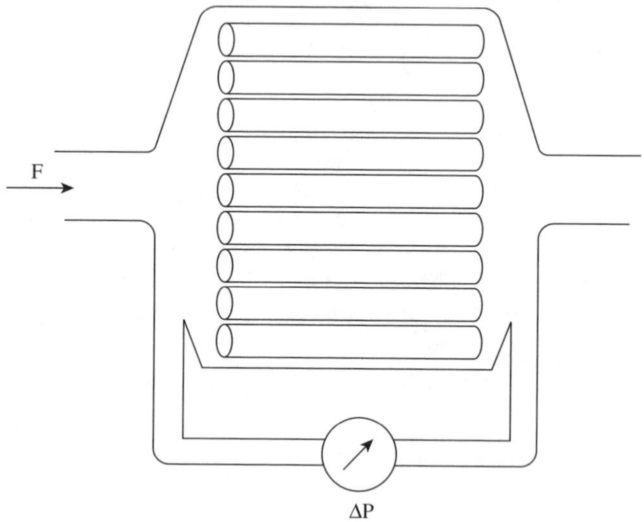

Figure 4.1. In the pneumotachograph, the total flow is divided between a large number of small parallel tubes. This guarantees a laminar flow and a constant resistance.

Temperature differences are prevented by heating the pneumotachograph assembly to 37°C (this can take several hours!). This is also done for another reason: water vapor, exhaled by the patient, would otherwise condense in the pneumotachograph's small tubes and block them, which would change the value of R. Condensation is impossible if the pneumotachograph's temperature is higher than the gas temperature.

Advantages of this transducer are its linearity even at high flow rates, its ability to measure bi-directional flows and its large bandwidth. The last is required to faithfully measure rapid flow changes. A disadvantage is its relatively large dead space, which makes the pneumotachograph less suitable for small patients (children).

4.1.2. Other obstructive differential pressure transducers

Variants of the pneumotachograph exist which avoid the problem of its large dead space. The *fixed orifice* devices present a constant obstruction to the flow such as a constriction or a wire mesh, the *variable orifice* devices a flow-dependent obstruction such as a movable flap; see Figure 4.2. With neither device is a laminar flow guaranteed, so the measurement is not linear, but it is sufficiently reproducible and the reading can be electronically linearized.

The advantage of variable orifice devices is that low flows generate a high ΔP and thus an accurate reading. Their disadvantage is a high R at low flows. At high flows, however, a variable orifice device may actually present less resistance to the flow than a fixed orifice device.

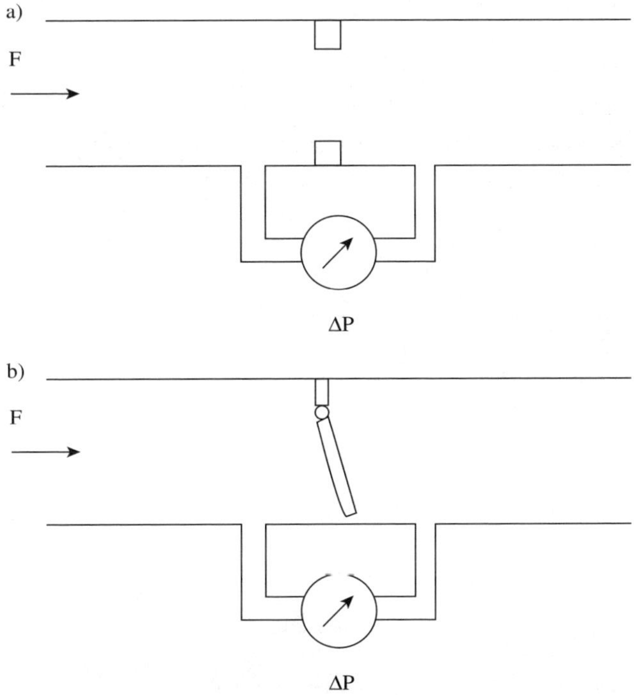

Figure 4.2. A fixed orifice flowmeter (a) has a constant obstruction. A variable orifice flowmeter (b) has a flow-dependent obstruction.

4.1.3. The rotameter

The rotameter consists of a vertical, slightly conical tube (Thorpe tube) with a float or bobbin in it. Gas flows upward and forces the bobbin up as well. Only unidirectional flows can be measured; attempting to reverse the flow would force the bobbin down, where it would block the tube. The bobbin's side surface has spiral grooves which make it rotate and keep it free from the walls of the tube (see Figure 4.3). Rotameters can only be used to measure *constant* gas flows; variations of the flow would make the bobbin, due to its mass, oscillate so much up and down that it would be impossible to record its height.

The height of the bobbin is determined by the equilibrium of two forces, a flow-induced force pushing the bobbin up which is equal to SΔP, where S is the bobbin's horizontal surface area and ΔP the pressure difference across the bobbin, and a gravity-induced force mg, where m is the bobbin's mass, which pulls the bobbin down. Thus, $\Delta P = mg/S$ is constant.

When the flow F is small, we can expect it to be laminar, so Poiseuille's law applies

$$\Delta P \sim \eta \, F / A^2 \tag{4.1}$$

or

$$A \sim F^{1/2} \tag{4.2}$$

where A now stands for the area through which the flow F passes around the bobbin. When the flow is large, we may expect it to be fully turbulent, so

$$\Delta P \sim \rho \, v^2 = \rho \, F^2 / A^2 \tag{4.3}$$

or

$$A \sim F \tag{4.4}$$

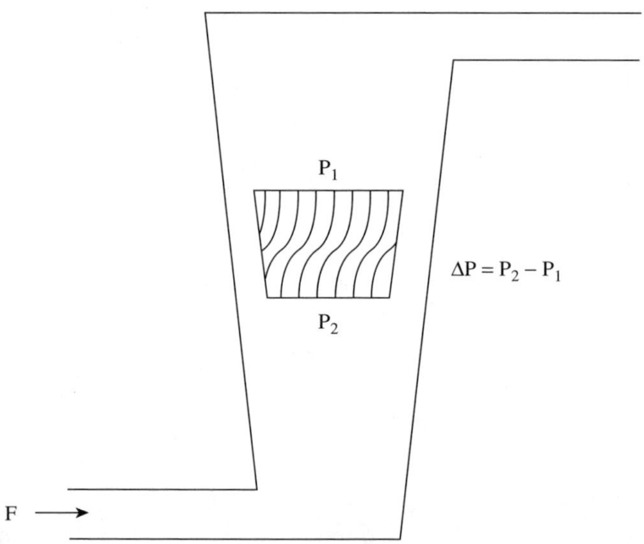

Figure 4.3. In a rotameter, a bobbin floats in a slightly conical tube. The height of the bobbin indicates the flow rate.

At intermediate flows we expect a relationship between A and F that is somewhere in between these two expressions.

Now let us compute the relationship between A, the area through which the gas flows around the bobbin, and the bobbin's height h. Since the rotameter tube is conical, its radius varies as

$$r\,(h) = R + c \tag{4.5}$$

where R is the bobbin's radius and c a constant. At height h, the tube's inner cross section is

$$\pi\,(R + c\,h) \tag{4.6}$$

of which an area πR^2 is blocked by the bobbin. The expression for A is thus

$$A(h) = \pi (R + c\,h)^2 - \pi R^2 = \pi (2\,c\,h\,R + c^2 h^2) \tag{4.7}$$

from which we can determine h (A) and thus also F as a function of h.

This discussion shows that the bobbin's height is related to the flow in a very complex way, which depends not only on the changing relationship between A (and h) and F, but also on the flow dependent importance of the viscosity and density of the gas (mixture) that flows through the rotameter. Viscosity and density of a gas mixture are strongly dependent on temperature, pressure and gas composition. A rotameter can thus only be used for a certain gas (if it is a mixture of gases, it must be of constant composition) at a specific pressure and temperature. Thus, a variety of rotameters are manufactured for different gases and different operating conditions; a color code denotes the gas (air: yellow; oxygen: green; nitrous oxide: blue). Using a rotameter that is calibrated for a certain gas for a different gas will generally give highly erroneous readings.

Several rotameters are often incorporated in a "rotameter block" in ventilators and anesthesia machines (see Section 8.2.1), where they measure the individual flows of the gases that the patient can be provided with (air, oxygen, nitrous oxide, sometimes carbon dioxide). In this application, the gas (composition) that is measured by each rotameter is always the same, and the flow is constant and unidirectional.

Since the rotameter's bobbin height depends on the pressure difference across the bobbin, its height will fluctuate if the ventilator or anesthesia machine whose gas it supplies somehow causes pressure fluctuations (and thus flow and gas density variations) at the rotameter's outflow side. In some cases, these oscillations of the bobbin's height would be so large that they preclude the use of rotameters as flow measurement devices.

4.1.4. The turbine flowmeter

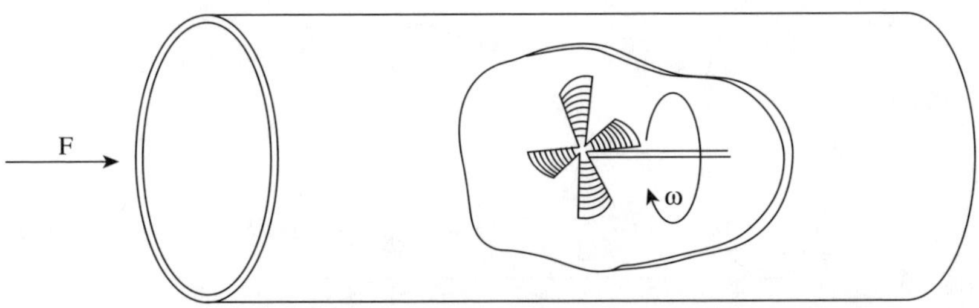

Figure 4.4. In a turbine flowmeter, the rotational speed of the turbine represents the flow.

A turbine flowmeter is depicted in Figure 4.4. A small propeller or turbine is brought into rotation by the flow. The turbine's angular velocity is a measure for the flow F passing through the meter. It is well suited to measure constant flows in either direction. A disadvantage is its low bandwidth, which is due to the turbine's inertia; it accelerates and decelerates slowly when the flow changes ("coasting"). It is therefore not well suited for time-varying bi-directional flows. If it is used, e.g., in the expiratory limb of a ventilator (see Section 8.2), it is often not the flow but its integral (i.e., tidal volume or minute volume) that is presented to the user. In the normal operating range, errors of up to 20% may be expected, but at low tidal volumes the error may be much larger.

Question: In a ventilated patient, the flow pattern accurately repeats from breath to breath. Show why the turbine's inertia is no problem if it is the tidal volume that must be determined.

4.1.5. The hot wire anemometer

In a hot wire anemometer (Figure 4.5), a known current I is sent through a small heater element in the flow, which will cool it. The wire is made of a material whose resistance is temperature dependent. The hot wire resistance R, whose value V/I can be determined by measuring the voltage across it, will thus depend on the flow F. The current I is made directly proportional to V; this can be done by a feedback system. Thus, R remains constant. A constant R implies a constant temperature T, effectively reducing the anemometer's thermal time constant to a very low value. This produces a large bandwidth, so that rapid flow changes can be measured. A disadvantage, however, is that no information is available about the direction of the flow.

The hot wire anemometer's reading depends on the amount of heat that the gas transports away from the wire, and this in turn depends on the temperature, pressure and composition (which determine density, viscosity, heat capacity and heat conduction coefficient) of the passing gas.

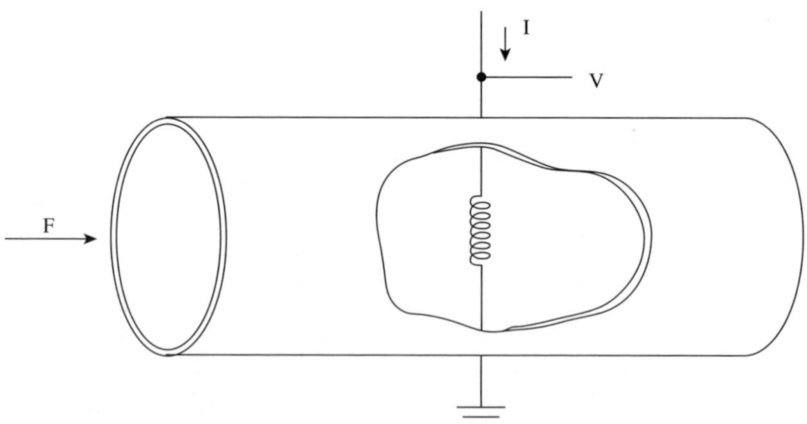

Figure 4.5. In a hot wire anemometer, the gas flow cools the small heating element. A feedback system, however, keeps its temperature constant. As a result, the electric current through the filament indicates the gas flow.

Because the hot wire can be made very small, the hot wire anemometer is an ideal instrument to probe the flow profile in a tube. For measuring the *total* flow its small size is a disadvantage, because errors will arise if the flow profile changes.

4.2. MEASURING BLOOD FLOW

The major problem with measuring blood flow is that we cannot — or do not want to — place an obstruction in the blood flow and measure the pressure difference across it. The techniques are therefore often limited to measuring blood velocity rather than flow. Flow can, of course, only be computed from velocities if 1) we can obtain the average of all velocities and 2) we know the cross section of the vessel. Both can be major problems.

4.2.1. The electromagnetic flowmeter

A conductor moving through a magnetic field generates an electric potential. This principle, also employed in, e.g., microphones and generators, is the basis for the electromagnetic blood velocity meter. This device does not measure the *flow* but the *velocity* of the blood. If we also know the cross section of the tube through which the blood flows, and if we also know the flow profile (or assume that it is parabolic), we can compute the blood flow.

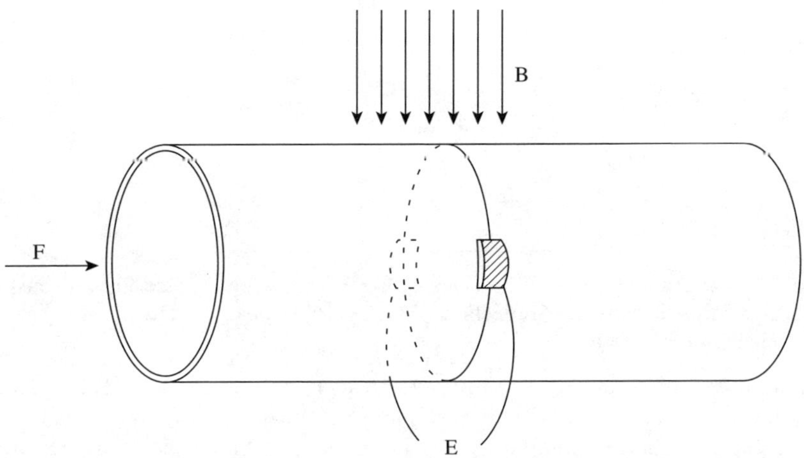

Figure 4.6. In the electromagnetic blood velocity meter, a conductor (blood) moving through a magnetic field B generates an electric potential E proportional to the average blood velocity.

In this application, the conductor is blood, the magnetic field B is externally applied, and the electric potential E is measured on the inside surface of the non-conducting, non-magnetic tube (Figure 4.6). E can be shown (the formulas are too complex to give here) to be the average of the velocities of the blood in the tube. The application can be expanded to cases where the vessel walls are conductive. The advantage is that now potential E can be measured on the *outside* of the vessel wall.

This technique is useful when the blood flow can be measured outside the body, e.g., in a heart-lung machine (see Section 8.3), which temporarily takes over the function of the heart. It is not practical for measuring the flow in intact blood vessels, where the major problem is that access to the vessel (and thus surgery) is required to position the inductor that generates B and the electrodes that measure E. Miniature devices have, however, been designed which can be put on the tip of a catheter.

4.2.2. Ultrasonic Doppler blood velocity and flow measurements

When moving objects reflect (ultra)sound, the reflection is frequency shifted. This is the well-known Doppler effect. Two different blood flow measurement techniques are based on this principle. In the *continuous ultrasound* technique, a single frequency is transmitted; in the *pulsed ultrasound* technique, a sequence of very short pulses is transmitted. In both techniques, the processing of the received signals results in the average blood velocity. If the vessel diameter and the flow profile are also known, the flow can be computed.

4.2.2.1. Continuous ultrasound

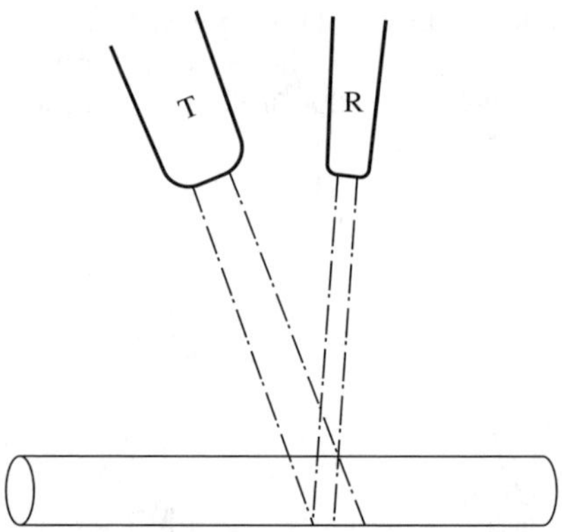

Figure 4.7. A transmitter generates a beam of ultrasound, which is partially reflected by the tissues. A receiver acquires the reflections. Moving objects cause a Doppler frequency shift proportional to their velocity.

When an ultrasonic transducer (a piezo-electric crystal) transmits a beam of a single frequency signal, the beam will be (partly) reflected wherever it encounters a change in acoustic impedance along its path. In particular, the signal will be scattered (partly reflected in all directions) by small objects in its path. The reflected signals are received by another ultrasonic transducer. If an object moves, the received frequency due to this object will show a Doppler frequency shift. In blood, there are a great many objects that move: the blood cells. Many different frequencies thus contribute to the received signal. The bandwidth of the received signal corresponds with the velocity range of the blood cells. If we assume that each cell's reflection contributes equal energy to the received signal, the power density at a certain frequency indicates the number of cells that move at the corresponding velocity.

Since the cells move with the blood, their velocities are the velocities of the blood itself.

Figure 4.7 shows the continuous ultrasound technique. In order to measure in one blood vessel only, the energy produced by the transmitter must be highly directional, and the signals received must constitute a narrow bundle as well. In order for scattering to take place, the wavelength of the transmitted signal must be much larger than the size of the blood cells. In practice, the transmitted frequency f_0 is several MHz and the Doppler shift is several kHz (Figure 4.8). The received frequencies are thus very close to f_0, but synchronous demodulation "subtracts" f_0 from the received signals and brings them down to low frequencies that can be easily processed.

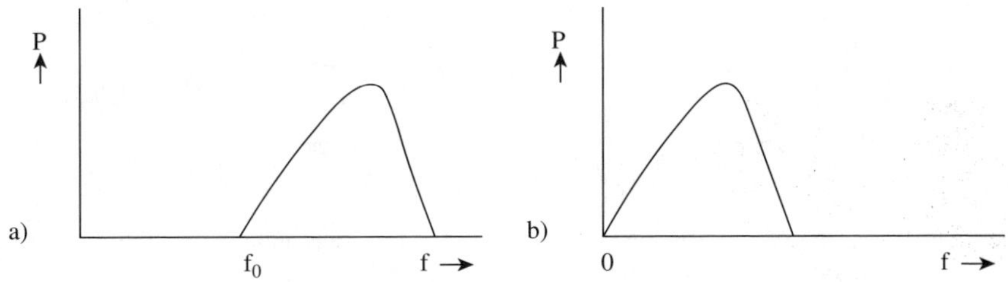

Figure 4.8. The received signal's frequencies vary around the transmitted frequency (a); a large echo at f_0 due to non-moving tissue is not shown. Synchronous demodulation subtracts the transmit frequency f_0, resulting in a low-frequency spectrum (b).

We will examine the scheme depicted in Figure 4.7 in some more detail. The signal is transmitted at frequency f_0 (which is identical to the mechanical resonance frequency of the crystal). A blood cell moving at velocity v will result in a received signal with frequency

$$f = f_0 (\cos\alpha + \cos\beta)\, v/c \tag{4.8}$$

where α is the angle between the transmitted beam and the axis of the blood vessel, β the angle between the "received beam" (the beam that is formed by the receiving characteristics of the receiver crystal; see Figure 4.7) and the vessel's axis, and c the sound velocity in the traversed tissues; the last is (almost) constant and has a value of 1500 m/s. Since f_0, α, β and c are fixed, f is linearly related to v.

Question: Assume that transmitter and receiver are built together into a single probe. Show how by maneuvering the probe one can find a position where the probe is perpendicular to the vessel.

The angles α and β and the cross sections of the transmitted and received beams determine the tissue volume from which we receive signals. Signals are received from all blood cells that have positions in the area where transmitter beam and receiver beam overlap, the *measurement volume* (the shaded area in Figure 4.7). If we assume that each blood cell contributes an identical power to the received signal, that the concentration of the cells in the measurement volume is homogeneous, that the flow is stationary and laminar and that the flow profile is parabolic, one can compute that the *average* velocity of all blood cells (and thus of the blood) is found by fitting a vertical line through the received spectrum so that the power to its left and the power to its right are equal (Figure 4.9). The found frequency f_{ave} indicates the average velocity v_{ave}. This result is relatively robust if the assumptions are relaxed. Deviations from a parabolic flow profile, for instance, cause only small errors. Major errors occur, however, if all velocities are not represented equally in the measurement volume; Figure 4.10 shows an example of this.

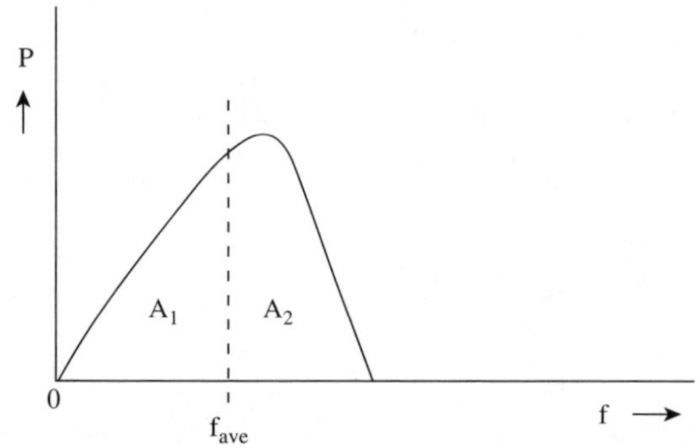

Figure 4.9. When the total received energy is divided into two equal parts, the dividing frequency f_{ave} is proportional to the average blood velocity v_{ave}.

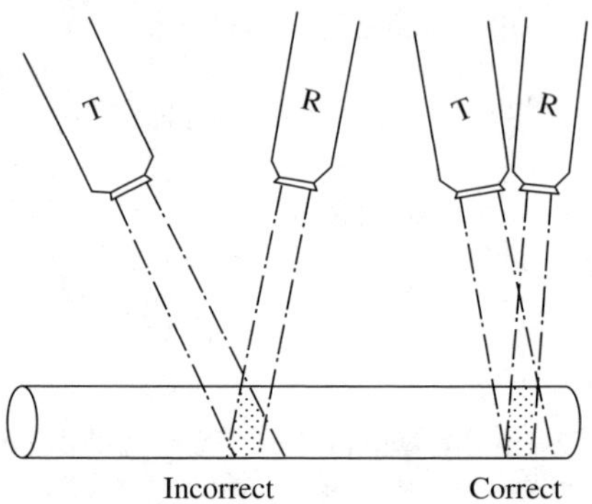

Figure 4.10. A measurement error arises when the measurement volume does not contain all blood velocities equally.

In practice, errors are also introduced by movements of other tissues than blood cells. High power low frequency components in the received signal are due to vessel wall motion, not to blood flow. These low frequencies need to be eliminated. This causes low blood velocities to be eliminated as well. Thus, those areas inside the vessel where blood flow is lower than a certain value (near the vessel walls) will not contribute to the results that the device presents to the user.

The transmitter and receiver transducers need to be placed on the patient's skin in such a way that no air will be found between transducers and skin; air greatly damps ultrasound transmission. In practice, a watery jelly is applied where the probe contacts the skin.

This technique measures blood velocity, not blood flow — except in those (few, artificial) cases where the vessel's lumen is known. The technique is very useful, however, to detect the positions of obstructions in vessels; where a vessel is very narrow, e.g., due to an atherosclerotic plaque, high velocities will exist.

4.2.2.2. Pulsed ultrasound

Whereas in the continuous ultrasound technique two transducers were required, a transmitter and a receiver, in the pulsed ultrasound technique one transducer suffices; this single crystal is alternately used as transmitter and as receiver.

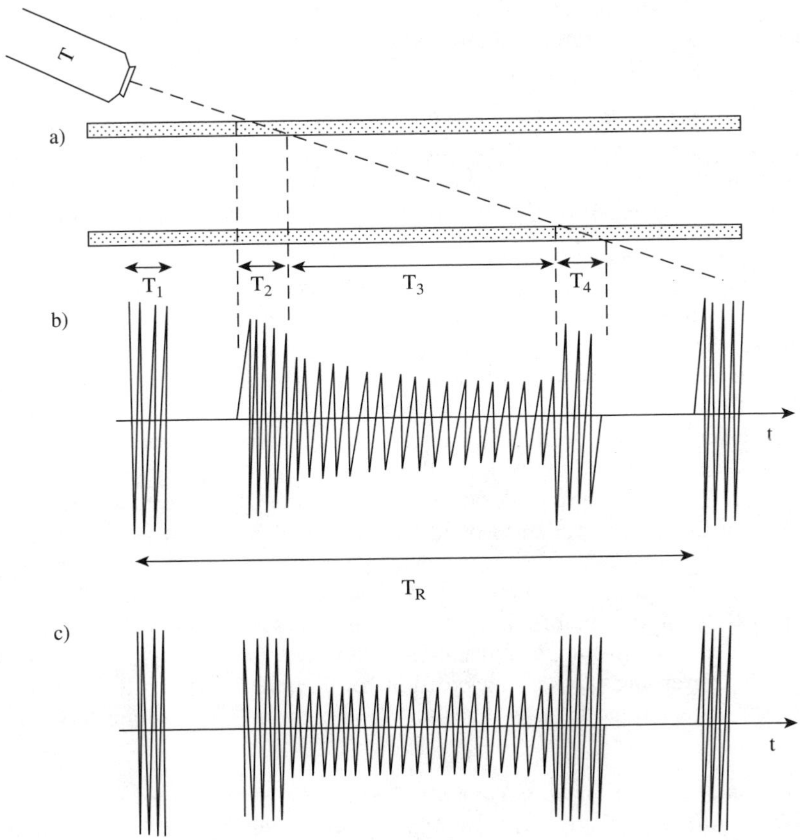

Figure 4.11. Echoes are generated when the beam of transmitter T encounters an acoustic impedance change (a). The time trace of the echoes is a one-dimensional "picture" of the tissues (b). Depth-dependent damping can be eliminated by a time-varying amplification (c).

The transducer, placed on the skin, transmits a short highly directional burst of ultrasound with a frequency of several MHz into the tissues of the body. The ultrasound is reflected wherever it encounters a change in acoustic impedance along its path. This echo can be directional if it is "mirrored" by a smooth surface (angle of reflection = angle of incoming beam), or it can be non-directional (scattering) by areas of differing acoustic impedance that are much smaller than the signal's wavelength. In biological tissues, no mirror-smooth surfaces exist; all echoes are (also) partly scatter reflections. In general, echoes will be received from all beam positions where a change in anatomical structure occurs. In this particular application, as indicated in Figure 4.11 where the transmitted beam has been idealized as very narrow, echoes are received from the vessel walls (T_2 and T_4) and from blood cells (T_3). T_1 is the time during which the transmitter is active, and the next pulse is transmitted after the repetition time T_R. Note that the power of the received signal decreases with the distance that the signal has traversed (back and forth); this is due to the absorption of ultrasound energy by the tissues. Since this absorption is approximately constant for all tissue types (except bones), it can be compensated for with a time-varying (exponentially increasing) amplification of the received signal (Figure 4.11c).

Question: Assuming that the ultrasound absorption is tissue-independent and thus purely depth-dependent, demonstrate that the amplification must increase exponentially from the time the pulse is transmitted until the time the next pulse is transmitted. Which problem(s) are necessarily introduced by the time-varying amplification?

The received signal is thus a one-dimensional "picture" of the anatomical structures that the transmitted beam traverses. Vessel walls, in particular, are easy to recognize (T_2 and T_4). The signal that is received during T_3 is due to the scattering by blood cells within the vessel. When these are in motion, the received signal is, due to the Doppler effect, shifted in frequency. This part of the signal can be isolated by detecting when the large reflections due to the vessel walls stop (start of T_3) and start again (end of T_3) and multiplying the received signal by zero except during this time ("gating"). The average blood velocity can then be determined in the same manner as in the continuous ultrasound technique. What is new is that there is now also information about the positions of the vessel walls, which can be used to estimate the vessel's diameter and thus its total flow. When the vessel is the initial segment of the aorta, before it branches, this technique provides a continuous readout of the cardiac output. What is also new is that vessel wall *motion* can be followed by observing the varying positions of T_2 and T_4. The latter is, in essence, the ultrasound technique for non-invasive blood pressure measurements (see Section 3.2.2.3).

Question: In order to be able to compute the vessel's diameter, the angle α between beam and velocity vector must be known. How can we obtain this angle by maneuvering the probe?

In order to compute the flow from average velocity and vessel diameter, all blood velocities must contribute equally to the received signal. If the beam is wide enough to enclose the whole vessel, this is ensured by processing the signal during the full period T_3. Instead of this, we can also process only a very short segment of T_3 ("gating"), and shift the position of this segment between the end of T_2 and the start of T_4. If we do this, we obtain the blood velocity at a particular depth in the vessel. By varying the depth, we can measure the velocity profile and thus the flow profile in the vessel. Since ultrasound is propagated at 1500 m/s, processing 1 μs of signal entails averaging over a depth of 1.5 mm.

The pulse repetition frequency must be chosen with care. All echoes must have been received before the next pulse can be transmitted; deeper lying tissues will generate echoes as well! If pulses are sent with a higher repetition frequency, it will be unknown to which depth a certain echo is related. If the time between pulses is T_R, the ultrasound has exactly this time at most to travel from transmitter to reflection site and back. During T_R, it travels a distance T_R c, where c is the ultrasound velocity in the tissues. Half of this distance is thus the maximum depth.

With a fixed T_R, the maximum depth is

$$d_{max} < \tfrac{1}{2}\, T_R\, c = \tfrac{1}{2}\, c / f_r \quad (4.9)$$

where f_r is the pulse repetition frequency. In practice, the maximum depth is of course given, and it determines the maximum pulse frequency

$$f_r < \tfrac{1}{2}\, c / d_{max} \quad (4.10)$$

Taking a worst case d_{max} of 50 cm and c = 1500 m/s, we find a maximum allowed f_r of 1500 Hz. Since, however, the Doppler spectrum in arteries can have a bandwidth of up to 6 kHz, Shannon's sampling theorem tells us that we should sample with a repetition frequency of at least 12 kHz. If, for this reason, we select f_r = 15 kHz, we limit ourselves to d_{max} = 5 cm. Deeper lying tissues will now contribute "noise" that we will have to live with.

A (very artificial) example will demonstrate the effect. Assume that pulses are transmitted every 70 μs, that there are equally well reflecting structures at depths of 1.6, 3, 15 and 30 cm, and that the beam is perpendicular to these structures. Echoes will arrive after 20, 40, 200 and 400 μs after the first pulse. After 70 μs, a new pulse is transmitted, and its echoes arrive at 90, 110, 270 and 470 μs after the first pulse was transmitted. Similarly, echoes again arrive at 160, 180, 340 and 540 μs, at 230, 250, 410 and 610 μs, etc. Arrival time ranges due to successive pulses overlap, and it is not clear anymore to which pulse an echo belongs. Analysis shows — and Shannon's theorem tells us — that a pulse received from a depth $d > d_{max}$, which therefore arrives after a time $2\, d / c > T_R$, will be interpreted as an echo arriving after a time 2 d / c *mod* T_R coming from a depth d′ = d *mod* d_{max}. If T_R = 70 μs, echoes arriving from depths of 1.6, 3, 15 and 30 cm after 20, 40, 200 and 400 μs will be interpreted as coming after 20, 40, 60 and 50 μs and thus from depths of 1.6, 3, 4.5 and 3.75 cm. Since tissue damping is proportional to depth, the amplitude ratio is 1:0.5:0.1:0.05. Thus, although deep tissues will contribute echoes, their amplitudes will be small.

Question: Earlier we introduced a time-varying amplification that compensates for tissue damping. What will be its effect on the noise contributed by deeper lying tissues?

There is also a limiting *minimum* depth, due to the finite width of the transmitted pulse. Echoes cannot be received while transmission is in progress.

Question: What is this minimum depth if the transmitted pulse has a width of one microsecond?

Figure 4.11c shows a so-called A-mode scan, where the A stands for amplitude: the echo's amplitude is plotted. A B-mode scan, where B stands for brightness, is obtained by scaling the amplitude to a brightness (from black to white), resulting in the display of a single line with darker and brighter sections. By varying the transducer's angle with respect to the underlying tissues, a two-dimensional picture appears. This principle is used in the electrocardiograph.

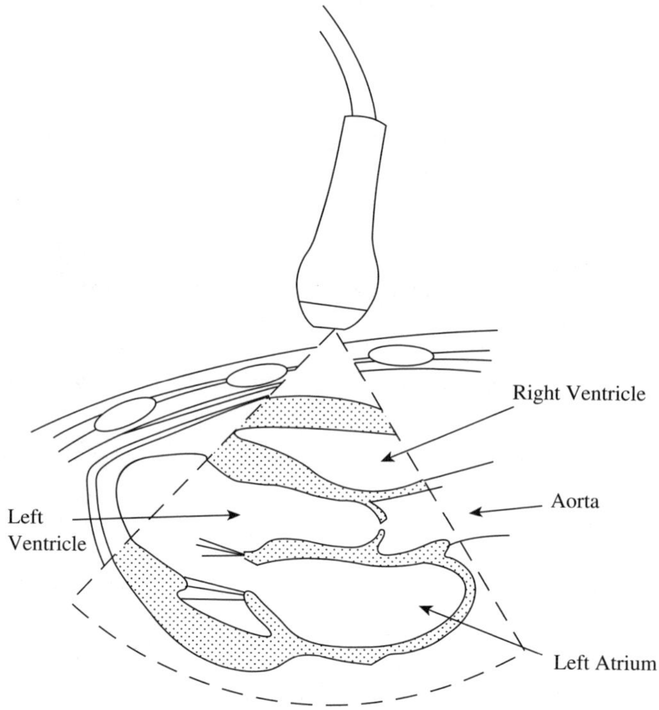

Figure 4.12. An echocardiogram of the heart.

In *echocardiography*, a multi-element ultrasonic transducer is used to generate a two-dimensional *image* of the heart. Such an image shows the anatomical structures of the heart and their movements, e.g., ventricular size and its change during ventricular contraction. The transducer probe can be placed on the chest above the heart or in the esophagus, close to the heart. The principle of the method is that acoustic pulses can be transmitted in different directions by offering slightly time-delayed electrical pulses to the individual elements of a one-dimensional array of transducers. The resulting wave front leaves the transducer at an angle that depends on the delay. The echoes that are received come from structures lying at this angle. By varying the delays, echoes can be collected from an almost 90° sector below the transducer. Figure 4.12 gives an example of an echocardiogram of the heart. The pulse repetition frequency is high enough to be able to follow movements of structures, e.g., heart valves, accurately.

QUESTIONS

1. Describe the pneumotachograph and how it can measure gas flow.
2. Discuss the relative merits of (a) fixed orifice and (b) variable orifice gas flow-meters.
3. Why is the height of a rotameter's bobbin related to the flow in such a complex way?
4. Why can a specific rotameter only be used for a specific gas?
5. Describe the hot wire anemometer.
6. Discuss how blood velocities are measured using the continuous ultrasound technique. When is this technique especially useful?
7. Discuss how the blood flow in a vessel can be measured using the pulsed ultrasound technique.
8. Which tissue depth limitations are there in the pulsed ultrasound technique? Why?
9. Explain how an acoustic pulse that is sent by a multi-element pulsed Doppler transmitter array can be given a varying spatial direction.

5 Measuring Composition and Concentration

The lung inhales and exhales respiratory gas. The composition of this gas mixture, i.e., the concentration of its components, can be measured by several techniques. Similarly, the composition of the blood, e.g., the partial pressures or the concentrations of the gases that the blood contains, can be measured. We will review the most often used techniques.

5.1. MEASURING GAS COMPOSITION

In order to determine the composition of a gas mixture — which individual gases it contains and how much of each — we somehow must differentiate between the gases. Since each gas has its own specific physical and chemical properties, this is always possible in principle. There are general methods, for example, mass spectroscopy, which allow a full decomposition into all components of interest. There are also specific methods, which determine how much of one particular gas is present in some mixture. These methods are based on a specific physical property of this but no other gas that might be present in the mixture. Oxygen, for example, is the only respiratory gas that is paramagnetic.

5.1.1. Mass spectroscopy

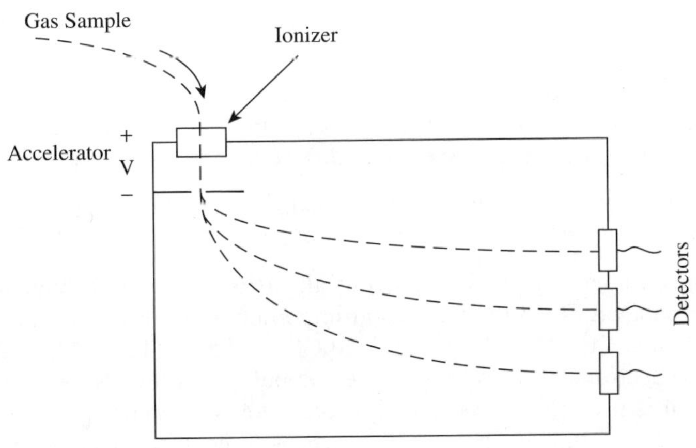

Figure 5.1. In a mass spectrometer, a small fraction of the gas to be analyzed is ionized. Ions are accelerated by an electric field and deflected by a magnetic field. Their position of impact on a detector depends on their mass and electric charge.

Figure 5.1 shows a schematic diagram of a mass spectrometer. A small flow of the gas to be analyzed is pumped to an ionization chamber, where the gas is ionized (electrons are removed from the gas molecules) by bombarding the gas by a beam of high velocity electrons generated by a hot wire filament (cathode). For simplicity, we assume that ionization removes one electron from the molecule. Positive ions with mass m are accelerated to a velocity v by an electric field E to an energy $E = \frac{1}{2}mv^2$ electronvolt; the resulting velocity of the ion thus depends on its mass and is different for ions with different mass: $v = (2E/m)^{\frac{1}{2}}$. A grid functions as an electrostatic "lens," which focuses the ions toward a small hole in the anode. Through this hole the ions enter a chamber where a magnetic field B, perpendicular to the chamber, acts on it. The vacuum in the mass spectrometer's chamber prevents collisions with other molecules from disturbing the ion's path. The result is a vectorial Lorentz force F due to B

$$\mathbf{F} = e\,\mathbf{v} * \mathbf{B} \tag{5.1}$$

where e is the ion's charge (positive, equal to the charge of an electron). Thus, the ion will move in a circle whose radius depends on its velocity v and hence on its mass m. At the positions where ions hit the chamber's wall, detectors (electrodes) are positioned where the ions pick up their missing electrons and thus generate a current. This current is proportional to the number of ions arriving per second and hence to the concentration of this particular molecule in the original gas mixture.

Question: In practice, ionization sometimes removes more than one electron from a gas molecule. How will this influence the ion's path?

Several variants of this basic characterization exist. It is possible, for instance, to use only one detector and modulate the electric or the magnetic field, so that all ion masses will arrive at this single detector at different times.

Question: Show that an ion with mass m can be given a path with radius r by choosing an appropriate E or an appropriate B.

The mass spectrometer method is general; all different gas molecule (ion) masses can be discriminated. Regrettably, several important gases have the same molecular mass, e.g., N_2 and CO (2*14 = 12 + 16) and CO_2 and N_2O (12 + 2*16 = 2*14 + 16). Also, a mass spectrometer is too expensive and bulky to allow its use for one patient only. Where it is used, it is time-shared; usually, the respiratory gas of each patient in an intensive care unit is analyzed for two or three successive breaths. This, in turn, means that each individual patient can only be monitored part of the time. This will be acceptable if the number of patients monitored with one mass spectrometer is not too large. A major disadvantage of using one mass spectrometer for an entire intensive care unit (with maybe 24 patients!) is that failure of the instrument becomes a major calamity; a second mass spectrometer is thus highly recommended for backup. A minor disadvantage is that the time delay due to pumping the gas from patient to mass spectrometer through very long lines makes it more difficult to relate its readings to the readings of flow and pressure, which have no delay.

5.1.2. Infrared absorption spectroscopy

The molecules of many gases absorb electromagnetic radiation (such as light). Normally, infrared light is used in gas analysis. How much infrared light is absorbed depends on the light's wavelength, the concentration of the absorbing substance, the length of the path of the light through the absorbing medium, and the substance specific absorbing characteristics. *Beer's law* (also called the Lambert-Beer law or the Beer-Lambert law) states this relationship (Figure 5.2a)

$$P_d = P_0 \, e^{-alC} \tag{5.2}$$

where P_d is the detected power, P_0 is the ingoing power, a is a constant which depends on the absorbing medium and on the light's wavelength (the molar extinction coefficient), l is the optical path length and C is the concentration of the absorbing medium. Absorption by some medium in a test cell (constant l) irradiated by a constant light energy of a given wavelength (constant a and P_0), thus only depends on its concentration. Thus, concentration can be determined from absorption: $C \sim \log (P_d)$.

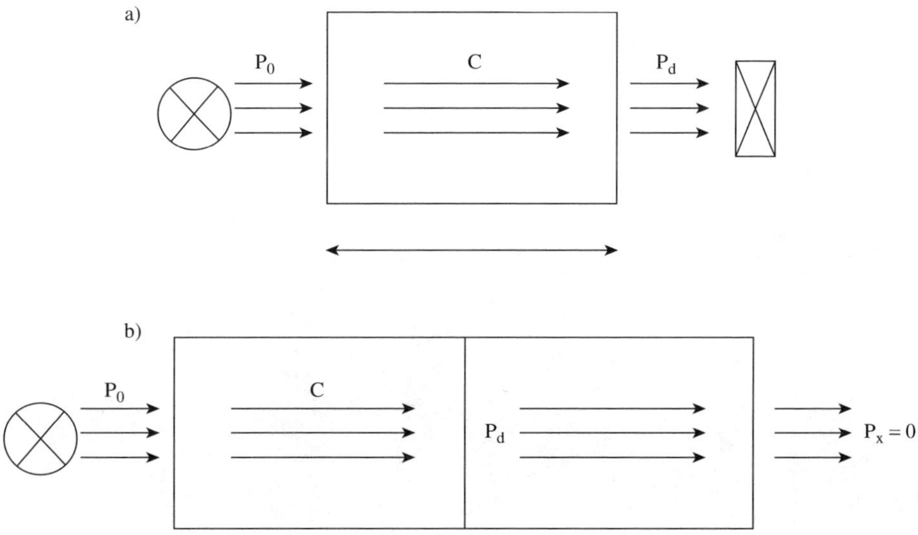

Figure 5.2. The energy P_0 of a light source is partly absorbed in the test chamber. A photodetector measures the remaining energy P_d (a). The gas itself can also absorb the remaining energy P_d (b).

Absorption of a photon by a gas molecule takes place when the oscillation, that normally takes place between the atoms of a molecule due to the elasticity of their chemical binding, has the same frequency as the incoming light. This is a quantum mechanical effect: only specific energies are able to change the energy level of a molecule; the photons, present in the incoming light, must have exactly the right energy in order to be absorbed. The effect takes place only when the molecule's electrical dipole momentum can change. This is normally the case for asymmetric molecules such as N_2O, H_2O, CO and CO_2, but not for symmetric molecules such as

O_2 and N_2. This is an inherent limitation of the method. Figure 5.3 shows the infrared absorption spectra of several gases. Clearly visible is that the spectra of some gases (e.g., CO and CO_2) partially overlap.

If a light source were available whose frequency could be tuned over the whole range of interest, the absorption peaks would indicate which gases were present in a gas mixture and how much of each. Since such broadly tunable light sources do not exist, current instruments employ separate light sources (usually infrared light-emitting diodes or LEDs, which can be manufactured for many different frequencies) at a fixed frequency that is carefully matched to the absorption peaks of a specific gas, to analyze the concentration of that particular gas. If the concentrations of more gases must be measured simultaneously, each gas requires its own light source, but multiple light sources may be mounted in a common test chamber. Medical infrared gas analyzers have been designed for CO_2 (0 to 10%), CO (0 to 0.3%), N_2O (0 to 100%) and several anesthesia gases. Their accuracy is in the order of 1% full-scale error, and a step response in the order of 100 ms is good enough for medical applications.

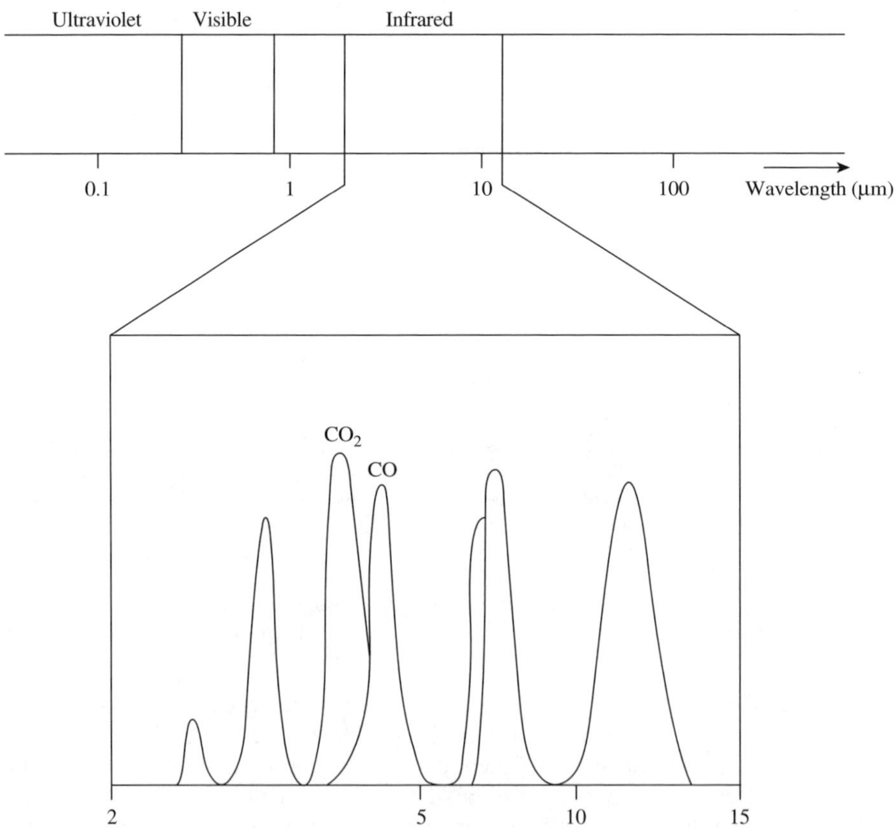

Figure 5.3. Different gases have different absorption spectra. Absorption spectra frequently overlap.

5.1.2.1. The capnograph

The instrument that measures the CO_2 concentration is called a *capnograph.* Several types of capnograph exist. The simplest type is the *in-line* capnograph, where a light source and a photodetector are mounted on opposite sides of a short transparent tube (Figure 5.4) through which the gas to be analyzed flows (an extension of the Y-piece, see Figure 1.15). The light source is a LED with a frequency corresponding to the peak of the CO_2 absorption spectrum. The photodetector must be able to detect light at this frequency with good efficiency, but its spectrum can be (much) wider, because no other light frequencies will be present in the received signal if the assembly is well shielded. Due to the frequency specificity of the light source, only the absorption due to CO_2 will be measured. However, if some CO is present as well (as is the case in heavy smokers), the measurement will be slightly inaccurate due to the overlap in absorption spectra.

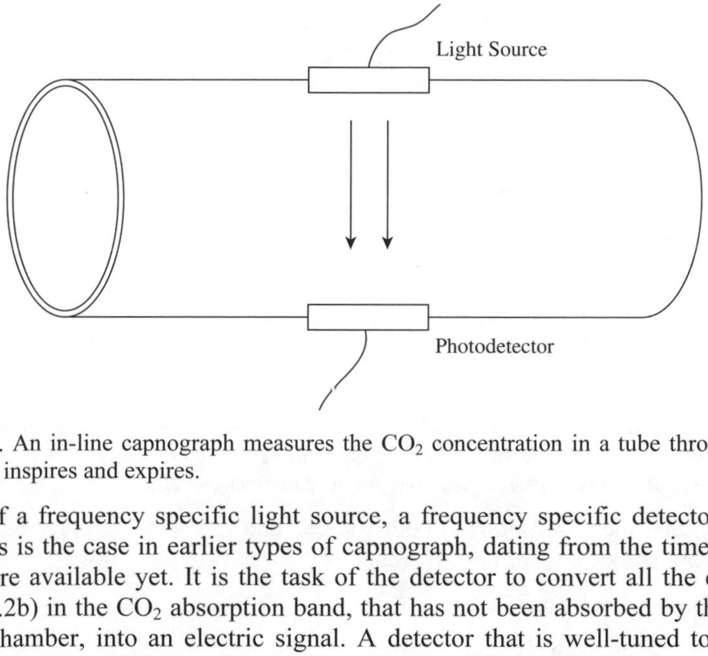

Figure 5.4. An in-line capnograph measures the CO_2 concentration in a tube through which the patient inspires and expires.

Instead of a frequency specific light source, a frequency specific detector may be used. This is the case in earlier types of capnograph, dating from the time when no LEDs were available yet. It is the task of the detector to convert all the energy P_d (Figure 5.2b) in the CO_2 absorption band, that has not been absorbed by the CO_2 in the test chamber, into an electric signal. A detector that is well-tuned to the CO_2 absorption band is of course CO_2 itself: a chamber filled with a high concentration of only CO_2 will absorb all incoming light at those frequencies where CO_2 absorbs, but at no other frequencies. Macroscopically, the detector will measure P_d (and thus the absorption in the test chamber) as a temperature increase — and a pressure increase, which can be measured.

Whereas in the in-line capnograph the sensor is brought to the expired gas, the *side-stream* capnograph of Figure 5.5 brings the gas to the sensor. It is based on the principle of Figure 5.2b. It uses a *differential* pressure measurement for improved sensitivity and accuracy. The test chamber is provided with a sample of the gas to be analyzed on CO_2 concentration; the normal measurement range is 0 to 10% CO_2.

The reference chamber is filled with a fixed CO_2 concentration (5%). The absorption in the test chamber varies with its CO_2 concentration; the absorption in the reference chamber is constant. The two CO_2 filled detectors (Golay cells) absorb the light energy that remains after passage of the light through both test and reference chamber. Between both detectors we find a thin flexible metal membrane. A very sensitive pressure transducer (a capacitor microphone) detects the membrane movement as a change of capacitance, which thus represents the pressure difference between both chambers. Since this pressure difference is due only to the difference in the light energy in the CO_2 absorption bands which reaches the detector's chambers, the detector measures differences in CO_2 concentration between test and reference chamber.

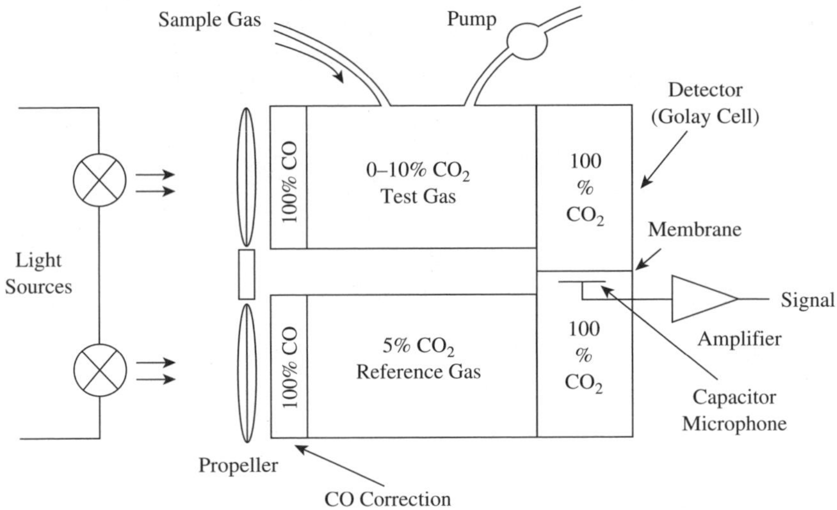

Figure 5.5. In a side-stream capnograph, some of the gas to be analyzed is pumped through a measurement chamber. The difference in absorption between measurement chamber and reference chamber generates a measurable pressure difference in the detector.

Assuming that the absorptions in both test cell and reference cell are small, we can approximate Beer's law for both chambers as

$$P_{dt} = P_0\, e^{-alC(t)} \approx P_0\,(1 - a\,l\,C\,(t)) \qquad (5.3)$$

$$P_{dr} = P_0\, e^{-alCr} \approx P_0\,(1 - a\,l\,C_r) \qquad (5.4)$$

where P_{dt} and P_{dr} are the energies arriving at and absorbed in test detector and reference detector, C (t) is the time-varying CO_2 concentration in the test cell, and C_r is the CO_2 concentration in the reference cell. We have also used the equality $e^x \approx 1 + x$. Since the flexible membrane's position, and thus the device's output signal S (t), depends on the difference between P_{dr} and P_{dt}, we have

$$S\,(t) = P_{dr} - P_{dt} = P_0\, a\, l\, (C\,(t) - C_r) \qquad (5.5)$$

A refinement is a mechanical chopper (much like the blades of a common ventilator), which alternates the light through both chambers. As a result of chopping, the capacitor microphone's pulsating signal can now be amplified by AC amplifiers, which are more stable. Thus, offset problems are avoided.

Overlap of spectra of different gases can be accommodated by passing only that part of the infrared light spectrum through the capnograph's cells that can be absorbed by CO_2 only. Since the spectra of CO_2 and CO overlap, light that could be absorbed by CO must not enter the test chamber. This can be done by adding an extra CO-filled cell in the light path, before the light enters the test chamber. This cell, which contains a high CO concentration, fully absorbs the overlap frequencies, so that any CO that might be present in the test gas can no longer absorb those frequencies and thus does not contribute to the instrument's readout.

Since the CO_2 concentrations, and hence the absorptions, in test and reference chamber are almost the same, this method depends critically on how well the instrument ensures that pressure varies with CO_2 absorption only. Temperature fluctuations are eliminated by using a thermostat, which keeps the capnograph assembly at a fixed temperature (it may take several hours, however, before the instrument is at that temperature and ready for use). Pressure variations in the test gas, which are reflected in the gas densities, are usually measured and compensated for.

A disadvantage of side-stream capnographs is the fact that some gas must be pumped away from the patient. Although this flow may be low (0.5 to 1 l/min), it may be in the same order as the minute volume of babies or premature infants. The danger is then that the patient's lungs will be forcefully emptied. Returning the analyzed gas can eliminate this problem; this is possible since it has not been changed in any way. Another slight problem is the several seconds delay of the capnogram, due to the transport of the test gas from the patient to the instrument, with respect to pressure and flow measurements. If relationships between respiratory signals are to be analyzed (e.g., the CO_2 concentration as a function of expired volume, see Figure 6.6), it must be ensured that this delay is compensated for.

A serious problem in monochromatic (single wavelength) analyzers is encountered in anesthesia, where one of several different anesthesia gases may be used. The absorption spectra of these gases are not identical, but they overlap. A selector switch on these analyzers must therefore be adjusted according to the gas that is to be analyzed. An erroneous switch setting may cause fatal accidents. When, for instance, halothane is used but the switch is set to isoflurane, the display may indicate 1% "isoflurane" (which would be normal), whereas in reality 6% halothane is given (2% halothane may be lethal!). Polychromatic infrared analyzers are safer because they detect which agent is used.

5.1.3. Photoacoustic spectroscopy

Photoacoustic spectroscopy is a variant of (chopped) infrared spectroscopy. Again, infrared radiation, the wavelength of which is chosen to coincide with an absorption peak of a specific gas, expands the gas, which increases the pressure in the measurement chamber. This pressure increase is measured in the measurement chamber; there is no separate detector. A rapidly spinning wheel modulates the incoming light intensity sinusoidally at an audio frequency. As a result, absorption, temperature and pressure also vary sinusoidally. The pressure variation is thus *sound* at the modulation frequency, which can be picked up by a sensitive microphone, filtered, amplified and presented as a gas concentration. This is similar to how the side-stream capnograph, described above, works, except that its modulation was not necessarily sinusoidal. The novelty of photoacoustic spectroscopy is that more than one gas can be analyzed at the same time by modulating different light sources, each chosen for a particular gas, at different audio frequencies. These audio frequencies can be separated by tuned filters, and the output of each filter corresponds to the concentration of one gas.

5.1.4. The oxygen analyzer

Oxygen is the only respiratory gas that is paramagnetic. Its magnetic property can be exploited in a variety of ways. In one method, a strong magnetic field is modulated at an audio frequency. As a result, the oxygen in a gas mixture will periodically be attracted and concentrated by the magnetic field. This causes a pressure variation (sound), which can be picked up by a microphone. This method is therefore easily combined into one device with the photoacoustic method, which determines the concentrations of the other gases.

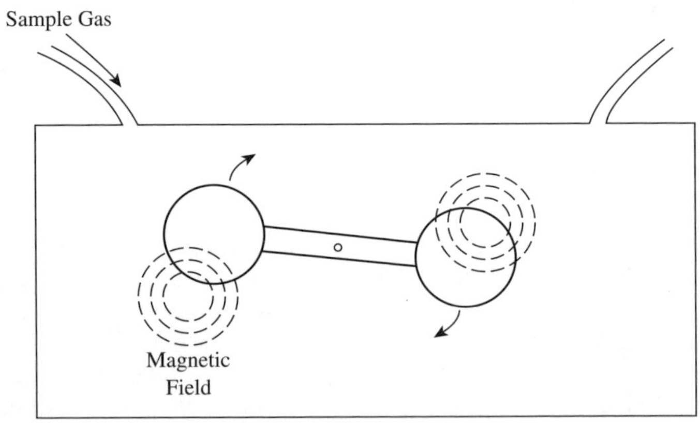

Figure 5.6. Oxygen is attracted to where the magnetic field is strongest, displacing the dumbbell balls from their positions. This displacement can be measured.

Another method also uses a strong magnetic field, but this field is not modulated. The field is, however, non-uniform. At the two positions where the field is strongest, lightweight glass balls (together forming a dumbbell) filled with a gas which does not contain oxygen, are affixed to a flexible spring wire, so that the dumbbell can easily rotate (Figure 5.6). When oxygen enters the measurement chamber, it is attracted to where the magnetic field is strongest. As a result, the glass balls are slightly displaced from their original positions. The rotation of the dumbbell can be measured and presented as the oxygen concentration.

5.2. MEASURING BLOOD COMPOSITION

Blood consists of different components. The percentage solid matter is its hematocrit value (Section 1.3.2). It is easily determined by separating solids and matter in a centrifuge. The red blood cells contain hemoglobin, which can carry oxygen. The percentage of hemoglobin that has bound oxygen is called the blood's *oxygen saturation*. Dissolved in the blood are a number of molecules and ions, and the blood gases, of which O_2 and CO_2 play the major roles. Hematocrit, oxygen saturation, blood gas partial pressures and blood electrolyte concentrations can all be determined from blood samples in a clinical laboratory. In a patient monitoring context, the focus is on *continuous* measurements, however. Only few real-time measurements of blood properties are in standard use.

5.2.1. Measuring oxygen saturation

The oxygen saturation of the blood denotes the percentage of the hemoglobin that carries oxygen, i.e., exists in the form of oxyhemoglobin.

5.2.1.1. The oximeter

Beer's law (see Section 5.1.2) can also be used in liquids such as blood. Different molecules will generally have different infrared absorption spectra. Oxyhemoglobin and hemoglobin are different molecules. Figure 5.7 shows their absorption spectra.

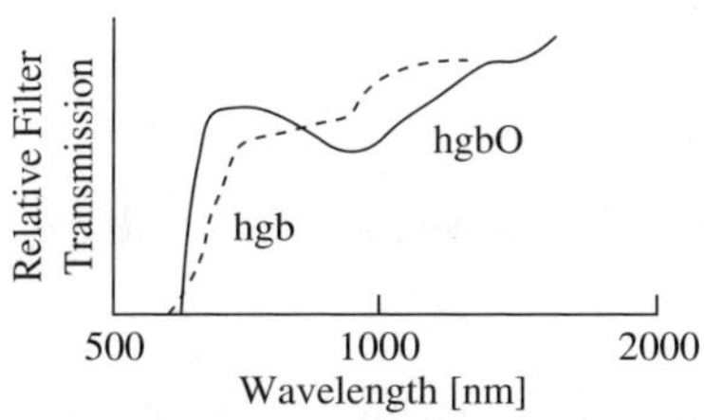

Figure 5.7. Hemoglobin and oxyhemoglobin have different absorption spectra.

Beer's law, whose expression is

$$P_d = P_0\, e^{-aLC} \tag{5.6}$$

can be rewritten to

$$P_d/P_0 = e^{-aLC} \tag{5.7}$$

which for aLC $\approx$ 0 (little absorption) can be approximated as

$$P_d/P_0 = 1 - aLC \tag{5.8}$$

or

$$A = 1 - P_d/P_0 = aLC \tag{5.9}$$

The term $1 - P_d/P_0$ is called the absorbance A. Noting the wavelength dependency of a and A and the fact that we have absorbance from both hemoglobin hgb and oxyhemoglobin hgbO, we can rewrite this expression as

$$A(\lambda) = a_o(\lambda)\, L\, C_o + a_r(\lambda)\, L\, C_r \tag{5.10}$$

where λ is the light's wavelength, L is the optical path length, a_o is the absorption due to hgbO, a_r is the absorption due to hgb, C_o is the concentration of hgbO, and C_r is the concentration of hgb.

We want, however, to know the *oxygen saturation*, that is, the percentage of the total amount of hemoglobin that exists in the form of oxyhemoglobin. We therefore convert the above formula to *relative* concentrations. C_o, the concentration of hgbO, can be expressed as the product of the density of hemoglobin and oxyhemoglobin combined ($w = C_o + C_r$) and the relative concentration (or percentage) of hemoglobin $C_o / (C_o + C_r)$. Thus, we replace C_o by $w\, C_o$ and C_r by $w\, C_r$. The term *relative concentration* means that $C_o + C_r = 1$. After this change of variables we have

$$A(\lambda) = L\, w\, [a_o(\lambda)\, C_o + a_r(\lambda)\, C_r] \tag{5.11}$$

where w is the density of hemoglobin and oxyhemoglobin combined and C_o and C_r are the relative concentrations of hgbO and hgb.

Now we perform a measurement at a wavelength λ_1 where $a_o = a_r = a$:

$$A(\lambda_1) = L\, w\, a(\lambda_1) \tag{5.12}$$

or

$$L\, w = A(\lambda_1) / a(\lambda_1) \tag{5.13}$$

Thus, the first measurement lets us determine the value of L w. From this follows

$$A(\lambda) = [A(\lambda_1) / a(\lambda_1)]\, [a_o(\lambda)\, C_o + a_r(\lambda)\, C_r] \tag{5.14}$$

Now we perform a second measurement at a frequency λ_2 at which a_o and a_r differ most. We get

$$A(\lambda_2) = [A(\lambda_1) / a(\lambda_1)]\, [a_o(\lambda_2)\, C_o + a_r(\lambda_2)\, C_r] \tag{5.15}$$

which, noting that $C_r = 1 - C_o$, can be rewritten as

$$C_o = k_1 + k_2\, A(\lambda_2) / A(\lambda_1) \tag{5.16}$$

where the two terms $k_1 = - a_r(\lambda_2) / [a_o(\lambda_2) - a_r(\lambda_2)]$ and $k_2 = a_r(\lambda_1) / [a_o(\lambda_2) - a_r(\lambda_2)]$ are constants that depend on known optical properties of the blood. In medical circles, the term C_o is called the *oxygen saturation* of the blood. The arterial oxygen saturation (the oxygen saturation of arterial blood) is written as S_{aO2}, the venous oxygen saturation as S_{vO2}.

Question: In the formulas above, one of the two wavelengths used was λ_1 where $a_o = a_r = a$, and the other wavelength was λ_2 where a_o and a_r differ most. Demonstrate that any two arbitrary wavelengths can be used, provided that the relative absorptions at those wavelengths obey certain properties.

The above formula works well for blood in a test tube. Tissues, however, consist of more than only blood. A modification of this method is therefore necessary if it is to become non-invasive and use the absorption of light in, e.g., a finger or an earlobe. Each tissue type has its own absorption characteristics. These must be compensated for, which requires measurements at additional wavelengths.

Oximeters are devices that can reliably measure oxygen saturation when light transmitter and sensor are placed on an extremity. They must, however, measure at many (e.g., eight) different wavelengths. This makes them expensive. Another disadvantage is that the light that passes through an earlobe is absorbed by both arterial and venous blood, generally in an unknown ratio. This makes the oximeter return a reading somewhere between S_{aO2} and S_{vO2}.

Catheters exist with a built-in oximeter. These oximeters use reflected rather than transmitted light and need only two light wavelengths. The light is conducted from an outside light source through fiber optics in a flotation catheter to the site whose saturation is wanted. This can be any place in the circulatory system that is accessible with a catheter. The light enters the blood, is reflected, and fiber optics carries it back to an outside photodetector. Both S_{aO2} and S_{vO2} can thus readily be obtained from different sites with errors of only a few percent.

5.2.1.2. The pulse oximeter

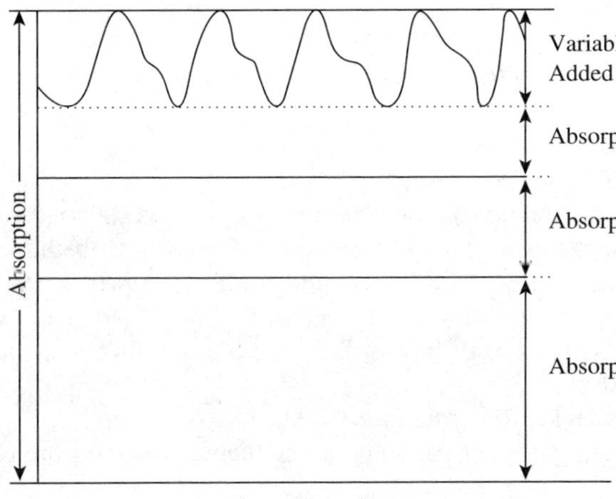

Figure 5.8. The absorption is due to different tissues. In a pulse oximeter, only the variation of the absorption is used, which corresponds with (part of the) arterial blood.

The oximeter became a very popular instrument when the next crucial step was taken, which reduced its cost and increased its utility. This step was to analyze the *changes* in absorption (Figure 5.8). These changes are due to the pulsations of the arteries, and changes in absorption must thus be completely due to arterial blood only. The absorption due to the tissues and the constantly present venous blood (and the part of the arterial blood that is also constantly present) is constant and disregarded. Measurements at two optical wavelengths suffice, as is the case for whole blood (660 and 940 nm are often used). This instrument is called the *pulse oximeter*. Its reading is often called S_{pO2} rather than S_{aO2}.

Pulse oximeters are valuable instruments that normally function well. Since pulse oximeters measure at two wavelengths, however, they can distinguish between two components only. They are meant to distinguish between hemoglobin and oxyhemoglobin, but blood can contain other hemoglobins as well, carboxyhemoglobin (hgbCO) and methemoglobin (Methgb), neither of which can carry oxygen. HgbCO levels, caused by smoking and urban pollution, are variable (up to around 10%) but can rise, directly after smoke inhalation, to 45%. Methgb levels also vary, up to 70%, above which death can occur.

We now consider the effect of Methgb on the pulse oximeter's readout; the effect of hgbCO is similar, but less extreme. In rare cases of the disease methemoglobinemia, which can be hereditary or acquired from the use of various drugs, the oximeter readout can be very inaccurate. In everybody, some of the hemoglobin is constantly being oxidized to methemoglobin, which cannot carry oxygen, but a normally occurring enzyme reduces methemoglobin back to hemoglobin. In case of a deficiency in the reducing enzyme activity, much of the hemoglobin may exist in the form of methemoglobin. Methemoglobin interferes with normal pulse oximeter function because it absorbs light almost equally well at the standard wavelengths of 660 and 940 nm.

QUESTIONS

1. Describe mass spectroscopy.
2. Describe infrared absorption spectroscopy.
3. How does a side-stream capnograph measure the carbon dioxide concentration in a gas mixture? How can overlapping spectra of different gases be handled?
4. Describe how, in the oximeter, Beer's law and the known properties of the absorption spectra of hemoglobin and oxyhemoglobin can be used to determine the oxygen saturation of blood. Why are measurements at (at least) two wavelengths necessary?
5. How does a pulse oximeter differ from an oximeter?
6. What is the effect of a significant presence of methemoglobin on the readout of a pulse oximeter?

6 Respiratory Measurements

The problem of how well the respiratory system functions is mainly approached through measurements of gas flow, gas pressure and gas composition and their derivatives.

Respiratory measurements are an important source of information in the lung function laboratory, where a suite of tests is available to test the various properties of a patient's respiratory system. These measurements provide a single static picture of the patient's condition, which is representative only if the properties of the respiratory system do not change (much) over time — i.e., if the condition is chronic. Respiratory measurements are also important in a critical care environment, where the patient's condition, which is critical and may therefore change from moment to moment, must be monitored continuously. These tests have a different character. Whereas in the lung function laboratory we can often count on the patient's cooperation, this is usually impossible in critically ill patients, who may have damaged lungs and who may be unconscious or otherwise unable to cooperate. For this reason the range of tests is less extensive, and also the reason that the tests must be performed (semi)-continuously.

6.1. MEASURING LUNG VOLUMES

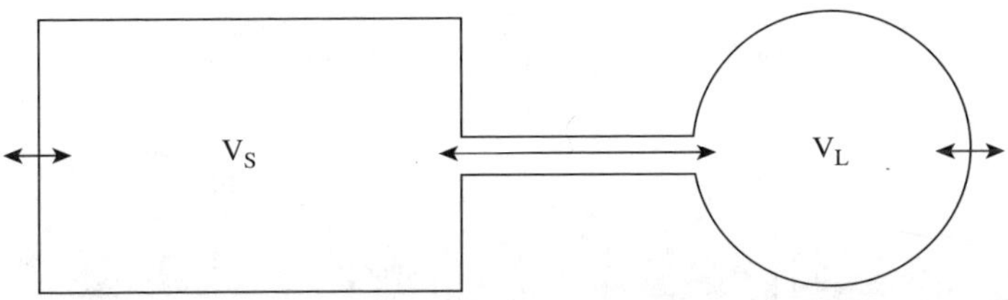

Figure 6.1. In spirometry, the patient breathes back and forth into the spirometer, whose volume V_s is measured. The pressure inside the spirometer is maintained at atmospheric pressure.

In the lung function laboratory, the various lung volumes are usually measured by having the patient breathe into and from a spirometer. The principle of spirometry is explained in Figure 6.1. The extra, highly compliant volume V_s of the spirometer is added to the lung volume V_l. Since the total volume $V_s + V_l$ is constant, changes in V_l will reflect changes of V_s, which can be measured. The spirometer is based on this principle. Figure 6.2 shows a schematic diagram. A lightweight airtight clock, whose mass is compensated for with a counterweight, can freely move up and down. A waterlock ensures that the movement encounters negligible resistance. The pressure inside the clock is thus atmospheric pressure. A carbon dioxide absorber ensures that patients will not rebreathe their expired CO_2. The CO_2 is chemically bound, e.g., according to

$$CO_2 + Ca(OH)_2 \Rightarrow CaCO_3 + H_2O \qquad (6.1)$$

In order to minimize the added dead space, *two* hoses, each with a unidirectional valve, connect the patient to the spirometer (see Section 1.2.7). A thermometer measures the temperature of the inspired gas; the expired gas is assumed to be at 37°C.

The various volumes that were defined in Section 1.2.5 can be measured by asking the patient to perform various maneuvers such as a maximum inspiration or expiration. Figure 6.3 shows a spirometer recording. Initially, the patient breathes quietly; volume changes reflect the tidal volume. Since the patient consumes oxygen, the spirometer's volume will decrease slowly over time. This decrease, which is visible as a slow drift superimposed on the volume changes due to normal breathing, is the patient's oxygen consumption. When the patient is asked to exhale and inhale maximally, the vital capacity can be read.

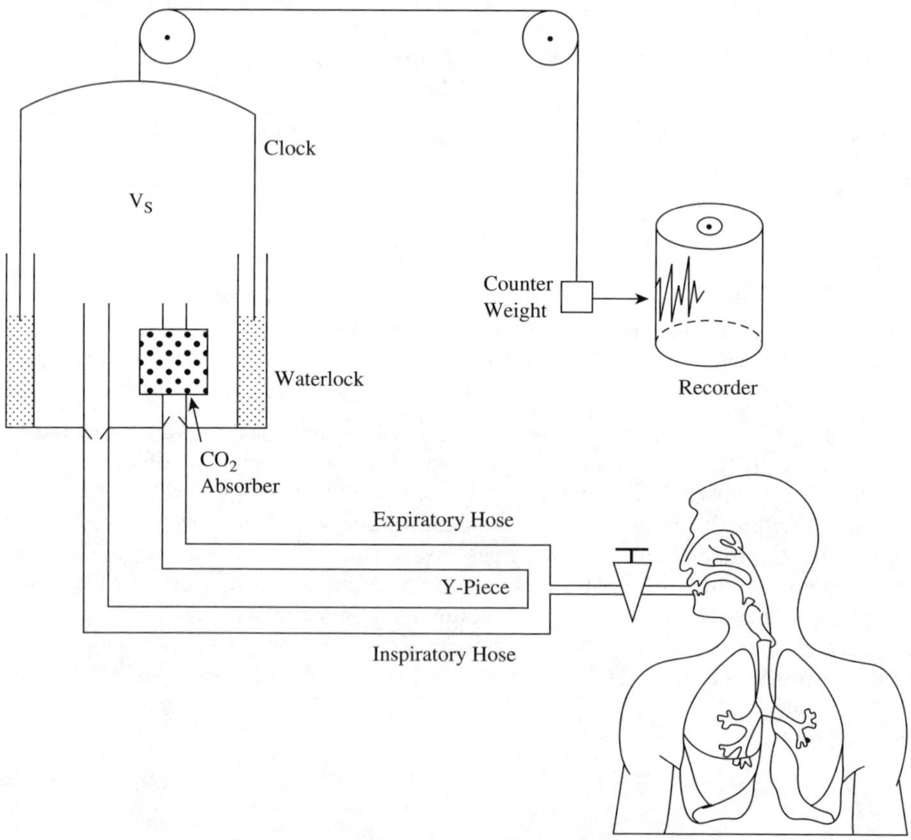

Figure 6.2. A spirometer is an airtight counterbalanced clock, whose volume is measured. Unidirectional valves ensure that minimal dead space is added. A CO_2 absorber removes CO_2 from the expiratory gas.

Figure 6.3 also shows some other maneuvers. The FEV_1 (Forced Expiratory Volume after 1 s) and the FIV_1 (Forced Inspiratory Volume after 1 s) provide measures for the power that the respiratory muscles can develop and for existing restrictions or obstructions. The MVV_{30} shows the tidal volume at a Maximum Voluntary Ventilation of 30 breaths per minute.

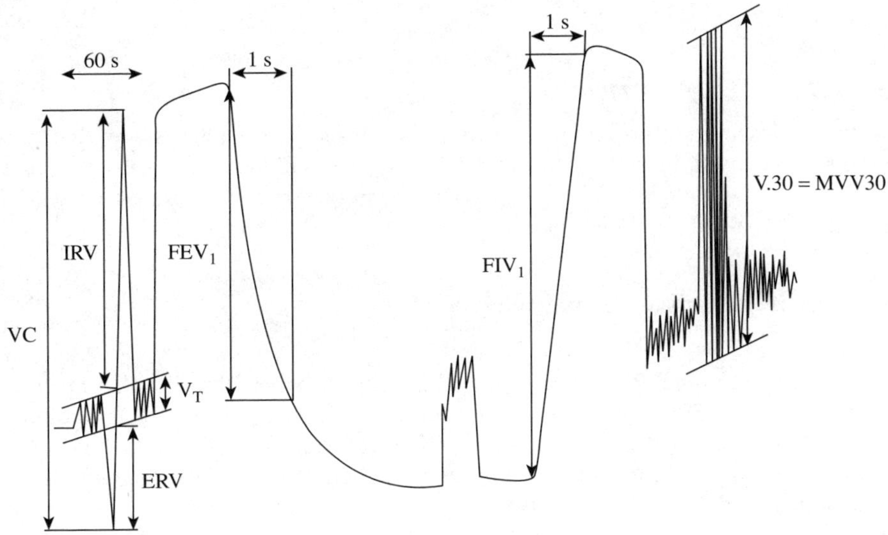

Figure 6.3. When the patient performs respiratory manoevers, the spirometer's volume changes reflect lung volume changes. The slow drift in the recording corresponds with oxygen consumption.

6.1.1. Corrections for temperature, pressure and water vapor

To accurately interpret and compare spirometry results, several corrections or compensations are necessary. When the gas moves back and forth between lungs and spirometer, its pressure, temperature and water vapor content vary. In slow maneuvers, the pressure inside the lung (the alveolar pressure) can be assumed to equal atmospheric pressure; the flow is small, and the pressure drop over the airway resistance is small too. In rapid maneuvers, where high flow rates occur, this is no longer true; in these cases, an esophageal balloon measurement (see Section 3.1.1) can establish the alveolar pressure, if required. The temperature of the gas in the spirometer is measured; the temperature of the gas in the lungs can be assumed to be 37°C. The expired gas is water vapor saturated. Some of this water vapor condenses on its way to the spirometer's lower temperature, but inside the spirometer the gas will still be water vapor saturated, but *at a different temperature*.

First, compensation for water vapor content is required. By subtracting the strongly temperature dependent water vapor partial pressure P_{H2O} (see Section 1.2.6) from the gas pressure P_{tot}, we obtain the "dry gas" partial pressure

$$P_{dry\text{-}gas}(T) = P_{tot}(T) - P_{H2O}(T) \tag{6.2}$$

We use Boyle's law (P V / T = constant) for the compensations for temperature and pressure. The result is the following formula, which converts a gas volume V′ from a temperature T′ and pressure P′ to an equivalent volume V at temperature T and pressure P

$$V(T) = V(T')\frac{P' - P_{H2O}(T')}{P - P_{H2O}(T)} \cdot \frac{T}{T'} \tag{6.3}$$

Question: Verify this formula.

In the following descriptions of spirometry tests, we will assume that these compensations will be performed. In the figures, we will also use the basic diagram of Figure 6.1 rather than the more complex Figure 6.2 in order to present the principles involved more clearly.

6.1.2. Measuring absolute lung volumes

The residual volume RV cannot be voluntarily exhaled; thus it, the functional residual capacity FRC and the total lung capacity TLC cannot be measured directly. The determination of these *absolute* lung volumes requires a different method. Three methods are in clinical use, the nitrogen washout method, the helium dilution method, and body plethysmography.

6.1.2.1. The nitrogen washout method

In the nitrogen washout method, the gas in the lungs — and thus the nitrogen in the lungs — is "washed out" into the spirometer (Figure 6.4). Initially, the spirometer is empty or contains a gas (mixture) in which no N_2 is present. During about 10 min or until the expired N_2 concentration is lower than 2%, the patient inhales pure oxygen and exhales into the spirometer. After this period, practically no N_2 is present in the lungs anymore. The reason is the alveolar membrane's low N_2 diffusion capacity; although a great deal of N_2 is in solution in the tissues, very little of it crosses into the lungs during a time as short as 10 min.

At the end of the test period, the N_2 concentration in the spirometer is measured. When the test starts, the normal concentration of N_2 in the alveoli is determined from the end-expiratory N_2 concentration; at the end of expiration, gas that comes from the alveoli is exhaled. Since the N_2 mass in the volume under consideration has not changed, we can compute the lung's total volume from

$$C_{N2}\,(t_{start})\,V_{lung} = C_{N2}\,(t_{end})\,V_{spir} \tag{6.4}$$

Question: Rewrite this formula including the corrections for water vapor partial pressure and temperature. Since the measurement normally starts and ends at the end of an expiration when the flow is approximately zero, no pressure correction is required.

If the FRC is to be measured (which is normally the case), the test starts and ends at the end of a normal expiration.

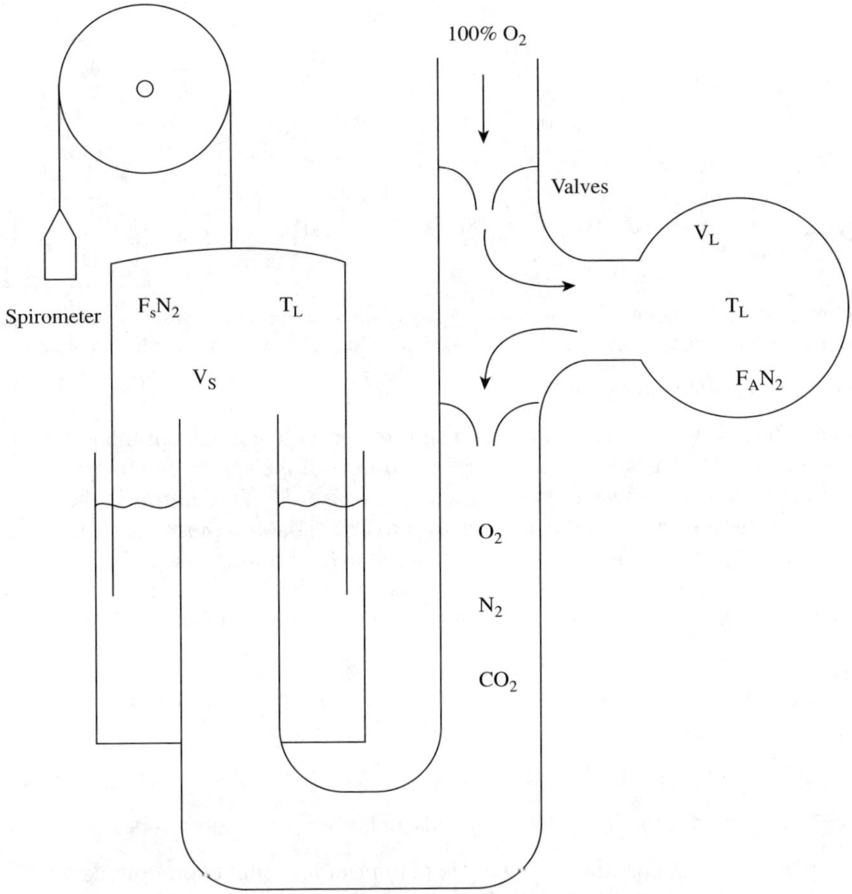

Figure 6.4. Inspiratory gas is 100% oxygen. During the test, the lung gases are expired into the spirometer. The final N_2 mass in the spirometer F_sN_2 is equal to the initial N_2 mass in the lung F_AN_2.

6.1.2.2. The helium dilution method

In the helium dilution method (Figure 6.5), the patient breathes back and forth into the spirometer. Initially, the spirometer contains a known (measured) concentration of helium. After a period of about 10 min or until the spirometer's He concentration does not change anymore, the helium has been redistributed over the larger volume of spirometer plus lungs combined and its concentration is measured again. Due to the alveolar membrane's low He diffusion capacity, practically no He has crossed from lung to blood and tissues during the test interval. Therefore, since the He mass has not changed, the relationship

$$C_{He}\,(t_{start})\,V_{spir} = C_{He}\,(t_{end})\,(V_{spir} + V_{lung}) \tag{6.5}$$

provides us with the total lung volume. Again, if the FRC is to be measured, the test starts and ends at the end of a normal expiration.

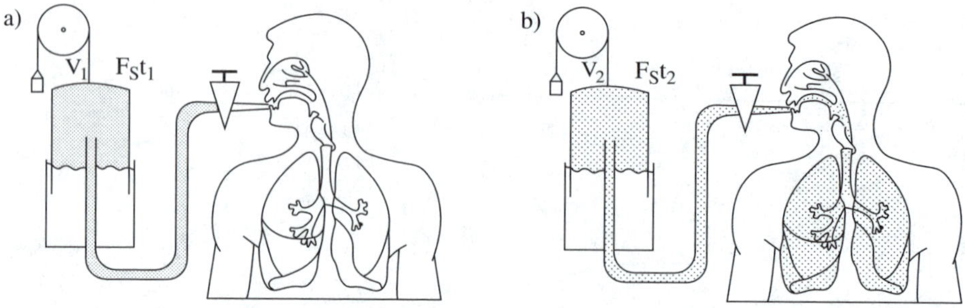

Figure 6.5. Initially, helium is found only in the spirometer and in the hoses. At the end of the test, it has been redistributed over a larger volume, which now includes the lung volume.

6.1.2.3. Body plethysmography

A body plethysmograph is essentially a spirometer large enough for the patient to sit in. Instead of the constant pressure varying volume of the spirometer, we now have a constant volume box whose pressure can be measured. To determine the absolute lung volumes, the patient inhales or exhales to a particular volume (usually FRC), at which time a shutter closes the tube through which breathing takes place. When the patient exerts inspiratory or expiratory effort against the closed shutter, the chest volume changes. The resulting pressure change in the box can be measured.

Applying Boyle's law (P V = constant), the chest (or lung) volume change ΔV_{lung} is computed. A second application of Boyle's law on the pressure and volume of the fixed amount of gas in the lung before and during the respiratory effort gives

$$V_{lung}(t_1)\, P_{mouth}(t_1) = V_{lung}(t_2)\, P_{mouth}(t_2) = (V_{lung}(t_1) + \Delta V_{lung})\, P_{mouth}(t_2) \qquad (6.6)$$

Since in this expression $V_{lung}\,(t_1)$ is the only unknown, its value can be computed.

If the patient's lung contains trapped gas volumes that cannot communicate with the respiratory tree, body plethymography and spirometry will give different results.

6.2. MEASURING DEAD SPACE AND UNEQUAL VENTILATION

Figure 6.6 shows the expired CO_2 concentration as a function of the expired gas volume. When an expiration starts, the first gas that is exhaled (region 1) comes from the anatomical dead space. This gas, inhaled in the previous breath, normally contains no CO_2. When the exhalation is well underway (region 3), the exhaled gas comes from the alveoli. In this region, the CO_2 concentration shows a slight increase. This *plateau slope* is due to the gravity-induced inhomogeneous emptying of the alveoli (see Section 2.10). In region 2, we observe a mixture of dead space gas and alveolar gas; as time progresses, the conducting zone contributes less and the respiratory zone contributes more to the expired gas. A numerical value for the dead space volume can be read from the curve as the volume where the CO_2 concentration reaches 50% of its final value, the end-expiratory CO_2 concentration.

The CO_2 concentration as a function of time is known as the capnogram. The expiratory part looks similar to Figure 6.6. During inspiration, the CO_2 concentration is normally zero. A period of a typical capnogram is shown in Figure 6.7.

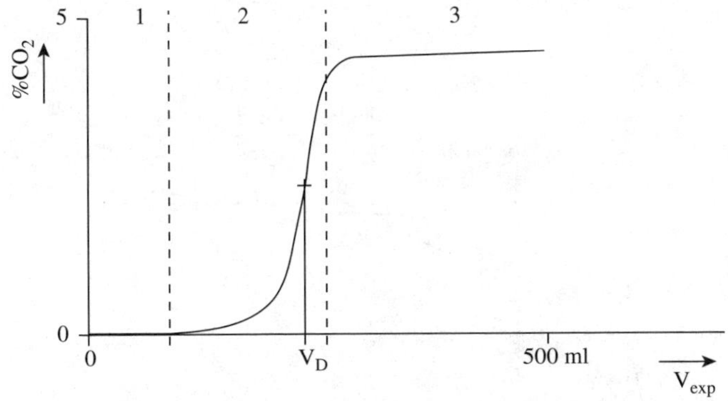

Figure 6.6. Expired CO_2 concentration as a function of expired volume.

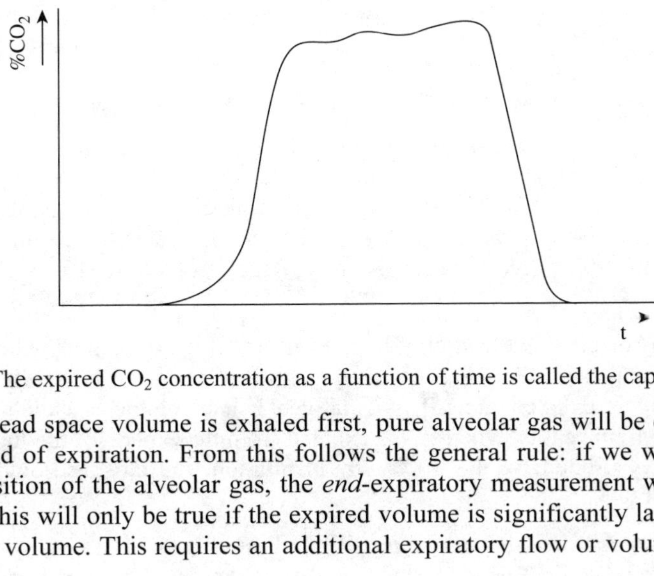

Figure 6.7. The expired CO_2 concentration as a function of time is called the capnogram.

Since the dead space volume is exhaled first, pure alveolar gas will be exhaled only near the end of expiration. From this follows the general rule: if we want to know the composition of the alveolar gas, the *end*-expiratory measurement will reflect it. However, this will only be true if the expired volume is significantly larger than the dead space volume. This requires an additional expiratory flow or volume measurement.

6.2.1. The single breath nitrogen washout method

The single breath nitrogen washout curve, shown in Figure 6.8, is similar to the CO_2 washout curve of Figure 6.6. It is obtained by measuring the N_2 concentration during a maximal expiration after a single maximal inspiration of 100% O_2. The inspiration starts at the residual volume. Initially, N_2-free dead air gas (pure O_2) is exhaled. Next comes mixed gas (partially dead space O_2, partially alveolar gas containing N_2). Then there is, as was the case with the CO_2 washout curve, a sloping plateau; its explanation is again that the larger alveoli at the top of the lung expand less; they take up less O_2, and thus the gas that they return is less "diluted" by O_2. Hence, it contains a higher concentration of N_2 (see Section 2.10). The curve of Figure 6.8 shows this.

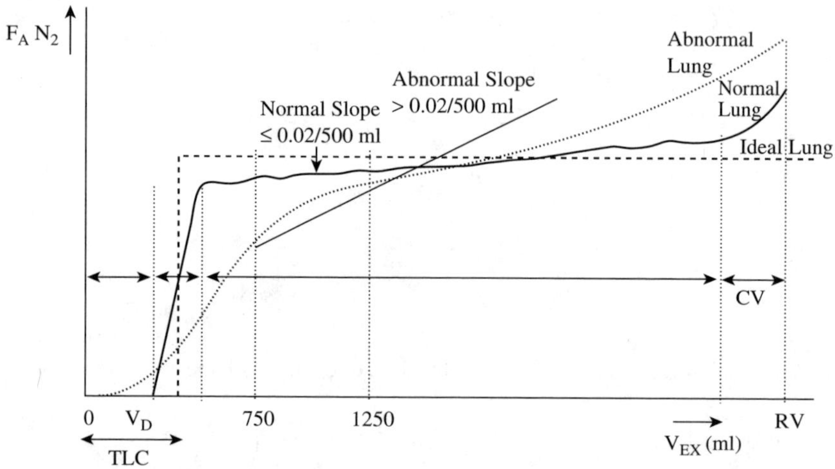

Figure 6.8. Expired N_2 concentration $F_A N_2$ as a function of expired volume.

The plateau concentration can be calculated if we know the residual volume RV and the total lung volume TLC. If we assume that RV = 1.2 l, it contains 0.8 * 1.2 l N_2, where the factor 0.8 is due to the fact that atmospheric gas contains 80% N_2. After inspiration of 100% O_2, this N_2 is assumed to be distributed over the whole lung volume. If TLC = 6 l, 0.96 l N_2 means an N_2 fraction of 0.96/6, i.e., a concentration of 16%.

The final part of the curve can only be seen in forceful exhalations, when the lower parts of the lung are fully collapsed. Only gas from the top of lung, with a large N_2 concentration, is exhaled here. This finally exhaled volume is called the *closing volume*. In extreme cases, when gas is exhaled from those parts of the lung that were not previously diluted by the 100% O_2 inspiration, the closing volume gas will contain about 80% N_2.

Figure 6.8 shows that the curve is different in abnormal lungs, which have a significant physiological dead space. Not only is the dead space volume larger, the mixing zone is wider as well. Unequal ventilation is also visible in a greater slope of the plateau. An index for unequal ventilation is obtained by fitting a straight line through the N_2 concentrations that occur after 750 and 1250 ml have been exhaled. Slopes larger than 0.02 (2%) per 500 ml are considered abnormal.

6.3. MEASURING LUNG COMPLIANCE

A simple way to measure the lung compliance is to have the patient breathe from a spirometer in volume increments of 500 ml while measuring the intraesophageal pressure. After each volume increase the lung needs a short no-flow stabilization period. It is essential that the patient's glottis remain open; only then will the alveolar pressure be zero. The measurement points, when connected by a smooth line, form a volume-pressure curve, as we already saw in Figure 1.6. The *static lung compliance* is the slope of the curve. Figure 6.9 shows a set of volume-pressure curves; the lung compliance (the slope of a curve) can vary a great deal, depending upon the physical properties of the lung tissues (normal, stiff or compliant lung). We also see that the curves are not straight lines. The lung compliance is not constant, although in the normal operating range it is nearly constant. In particular, the compliance decreases at high lung volumes. The figure also shows that the vital capacity VC is greatly decreased in both stiff and compliant lungs.

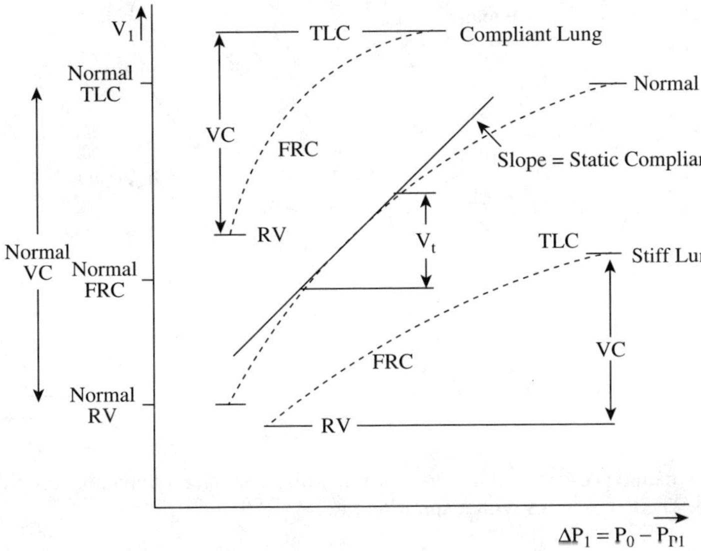

Figure 6.9. Static (no flow) pressure-volume curves of normal, stiff and compliant lungs.

The lung compliance can also be measured during breathing. When the patient breathes very slowly, the pressure drop over the airways is small and the alveolar pressure is almost zero, as long as the glottis is kept open. The resulting pressure-volume curve is the narrow loop of Figure 6.10. When a line is drawn through the points where the flow is zero (the end-expiratory and end-inspiratory points), its slope accurately reflects the static lung compliance.

When the patient breathes more rapidly, the PV-loop opens up and becomes wider. The compliance computed from such a curve is called the *dynamic lung compliance*. Only for very slow breathing is it equal to the static lung compliance.

If no intra-esophageal pressure is available, a *dynamic lung-thorax compliance* can be computed in spontaneously breathing or artificially respirated patients. Figure 6.9 showed that the lung's compliance decreases at large lung volumes. Lung overdistension occurs when the lung is forcefully filled (e.g., by a ventilator) to an excessive volume, where its compliance is small. This is visible in the PV-loop, which then has the shape of a duckbill (Figure 6.11). Such a curve signifies that little volume increases result in large pressure increases, which could cause barotrauma. A duckbill PV-loop indicates that the inspired tidal volume should be decreased; if the minute volume is to remain the same, the respiration frequency can be increased.

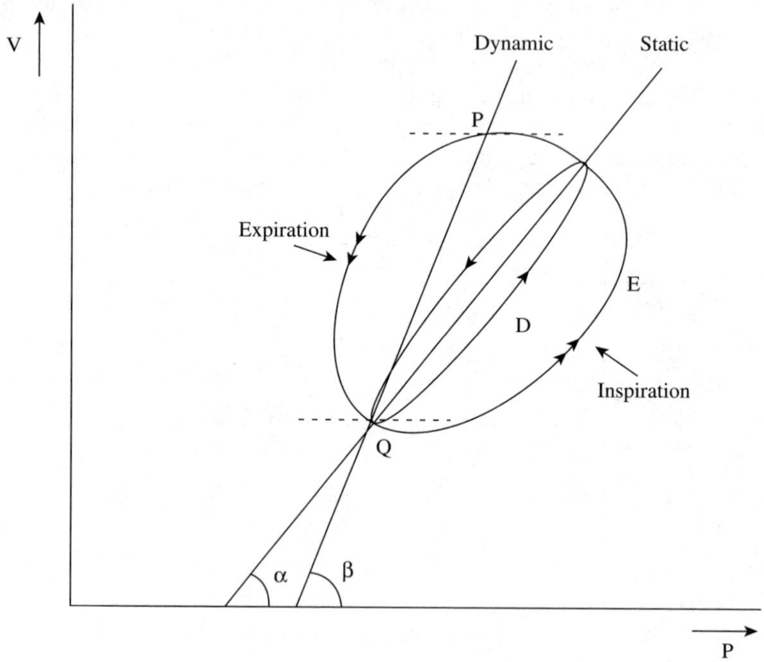

Figure 6.10. Dynamic pressure-volume loop, from which the lung's dynamic compliance can be determined. The latter varies with respiration rate.

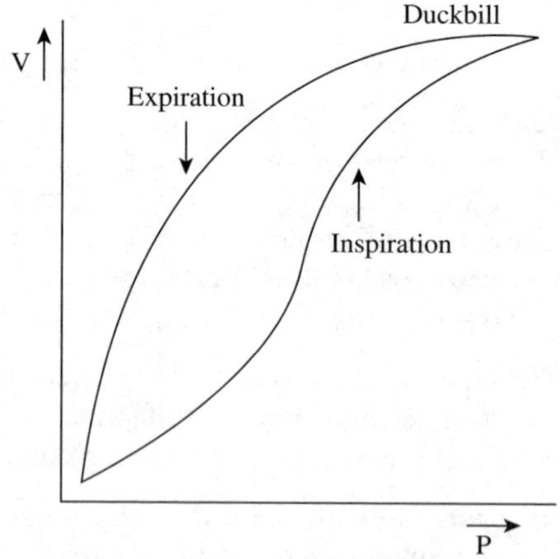

Figure 6.11. Lung overdistension results in a "duckbill" PV-loop.

Although the PV-loop is generally useful, a dynamic lung compliance computed from it may be unreliable, especially in constrictive lung diseases. Constricted air passages have a larger resistance; a model with a single resistance and a single compliance is now inadequate, since it assumes a uniform lung. An example will demonstrate this. Assume that the two lungs have equal compliances, but that the airway resistance of one lung is 10 times larger than that of the other lung (Figure 6.12a). Yet, our model is that of Figure 6.12b, which assumes *one* resistance R_{eq} and *one* compliance C_{eq}. At very low frequencies and flow rates, the pressure decrease due to the resistances is very small, and thus the dynamic compliance C_{eq} will be close to 2C. At higher frequencies, the pressure drop across the resistances becomes larger. Because one lung has an RC time constant 10 times larger than the other, it will take 10 times longer to fill it. If this time is not available due to the high respiration frequency, only one lung is functional and the dynamic compliance C_{eq} will approach the value C. Thus, the value of the apparent lung compliance will vary between C and 2C, depending on breathing rate.

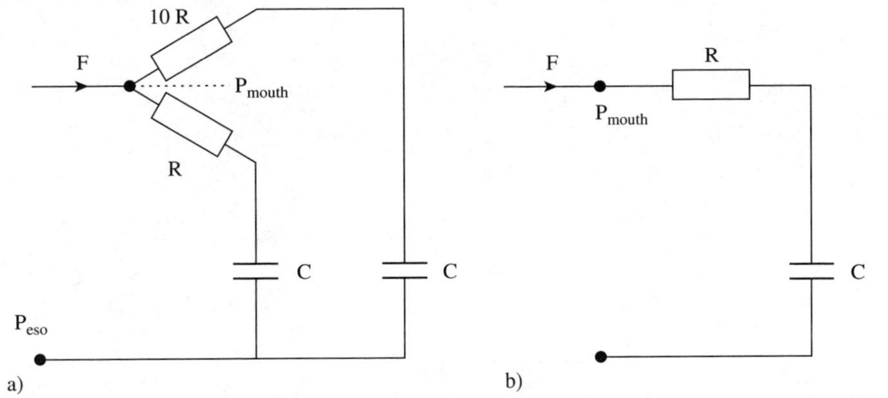

Figure 6.12. If the two lungs have different airway resistances and/or compliances, they must be modeled separately (a). The composite model on the right (b) is now incorrect: there are *two* time constants.

Question: Compute the value of C_{eq} of Figure 6.12b given the above assumptions, and show that it is frequency dependent.

This variability of the apparent dynamic lung compliance can be usefully employed as a test for unequal airway resistances. A curve of the apparent compliance, usually measured at a number of respiration frequencies between 10 and 120 per minute, will provide information about unequal constrictions.

6.4. FORCED EXPIRATORY FLOW-VOLUME CURVES

Plotting one variable as a function of another variable is a well-known technique that can bring additional information to light. We have seen this already in the pressure-volume loops. Figure 6.13 shows two flow-volume loops. The patient, who initially breathes normally (visible in the small loops at a volume of about 3.5 l), is asked to inspire maximally (to TLC) and then to fully exhale at maximum force (to RV). The curve is therefore called a maximum expiratory flow volume (MEFV) curve. When this is done twice, the curve reproduces only partially. Due to exhaustion or less exertion (the maneuver requires a great deal of effort), the peak expiratory flow (PEF) is smaller the second time. A well-reproducible value, however, is MEF_{50}, the maximally expired flow at the moment that 50% of the forced vital capacity has been exhaled. This value seems to be rather independent of how much force the patient exerts. The curve is also almost linear from the point MEF_{75}. This part of the curve is called the effort independent part of the MEFV curve.

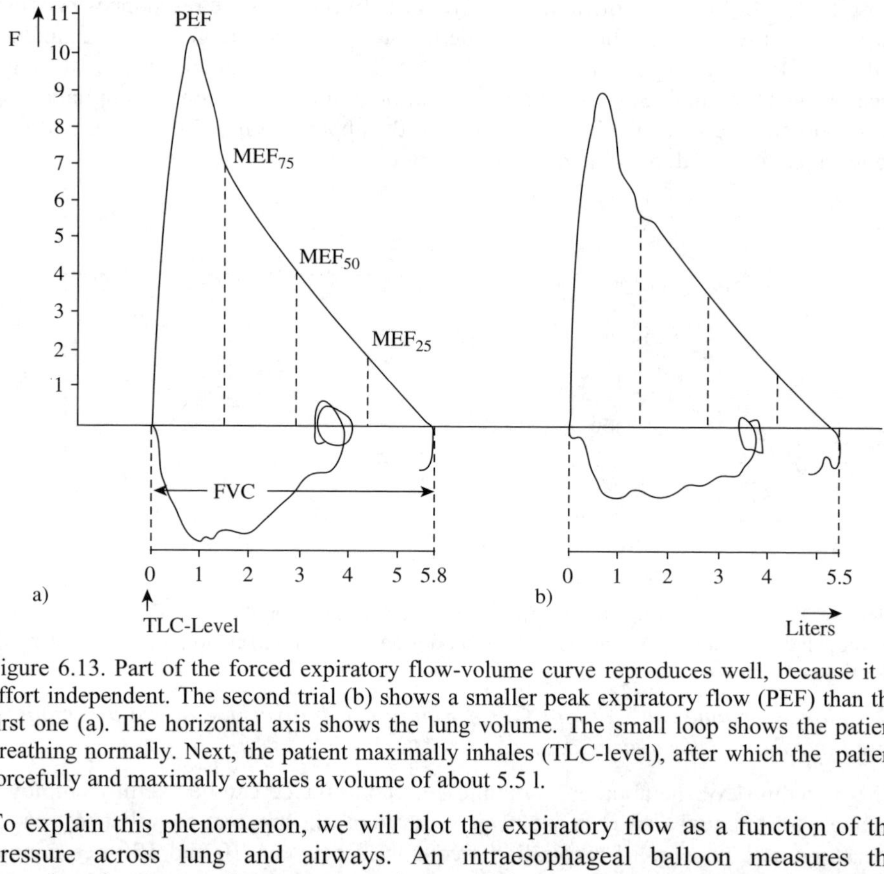

Figure 6.13. Part of the forced expiratory flow-volume curve reproduces well, because it is effort independent. The second trial (b) shows a smaller peak expiratory flow (PEF) than the first one (a). The horizontal axis shows the lung volume. The small loop shows the patient breathing normally. Next, the patient maximally inhales (TLC-level), after which the patient forcefully and maximally exhales a volume of about 5.5 l.

To explain this phenomenon, we will plot the expiratory flow as a function of the pressure across lung and airways. An intraesophageal balloon measures the intrapleural pressure; the pressure at the mouth is zero (with respect to atmospheric pressure). When we do this at different lung volumes, very different results are obtained (Figure 6.14). During inspiration (the lower left quadrant) there is a fairly linear relationship between pressure and flow; during inspiration, $C_{lung} = \Delta V/\Delta P$ is approximately constant. During forced expiration (the upper right quadrant), the flow tends to be volume dependent but pressure independent, especially at small lung volumes. Thus, during expiration C_{lung} appears to be approximately zero: a pressure change ΔP causes a negligible volume change ΔV. The explanation for this phenomenon has already been given in Figure 2.6 for a single collapsible tube: varying the pressure P_B outside the tube has no effect on the flow as long as the tube remains collapsed ($P_B > P_2$).

That only the lung volume determines the flow has been explained in Figure 2.7. Assume that with constant effort of the respiratory muscles the pressure across lung and airways is constant. In Figure 6.14 this could be represented as a vertical line at a certain intraesophageal pressure P_{eso}. During expiration, the operating point would then follow this line downward with a speed that continually decreases along its path. And that is exactly what we saw in Figure 6.13.

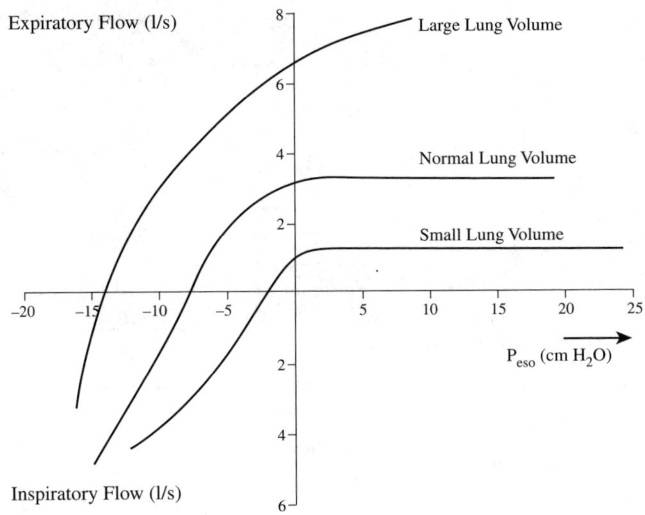

Figure 6.14. Expiratory flow as a function of lung transmural pressure at different lung volumes. At small and normal lung volumes, flow is effort independent.

A theoretical explanation has already been given in Section 2.9, where we saw that an externally applied pressure gradient field can modulate a flow through a large number of parallel small collapsible tubes. This is the case here. A *forced* expiration creates a high intrathoracic pressure. The pressure gradient field changes vertically due to the hydrostatic pressure; in an erect patient, the pressure is highest at the base of the lung (see Figure 2.10a). The pressure across an individual airway varies as well; this pressure is highest in the alveoli and zero at the mouth. Where the intrathoracic pressure is higher than the pressure in the airway, the airway collapses and its flow becomes zero. An exception arises in the large airways, whose anatomical structure prevents their collapse. Where the intrathoracic pressure is lower than the pressure in the airway, the airways do not collapse and contribute to the total flow.

Figure 6.15 shows this schematically for the whole lung. The left column presents the situation at the end of inspiration. At this zero flow moment, the pressure inside the alveoli P_A is zero. The pressure in the intrapleural space, measured with the intraesophageal balloon, depends on the lung's compliance and its volume according to $P = V/C$. The pressure gradient from alveoli to mouth is presented along the horizontal axis. Since there is no flow, the pressure is zero everywhere.

The right column shows the situation a fraction of a second later. What has changed is that the respiratory muscles have built up a pressure. The muscle action has two effects. The alveolar pressure P_A is raised, which will start to force the gas from the lung to the outside world; it creates a pressure gradient along the airways. In the figure, this gradient is approximated as a downward sloping straight line. The second effect is that the intrathoracic pressure is raised. In the figure, the intrathoracic pressure is approximated as constant everywhere along the airways and indicated by a dotted line.

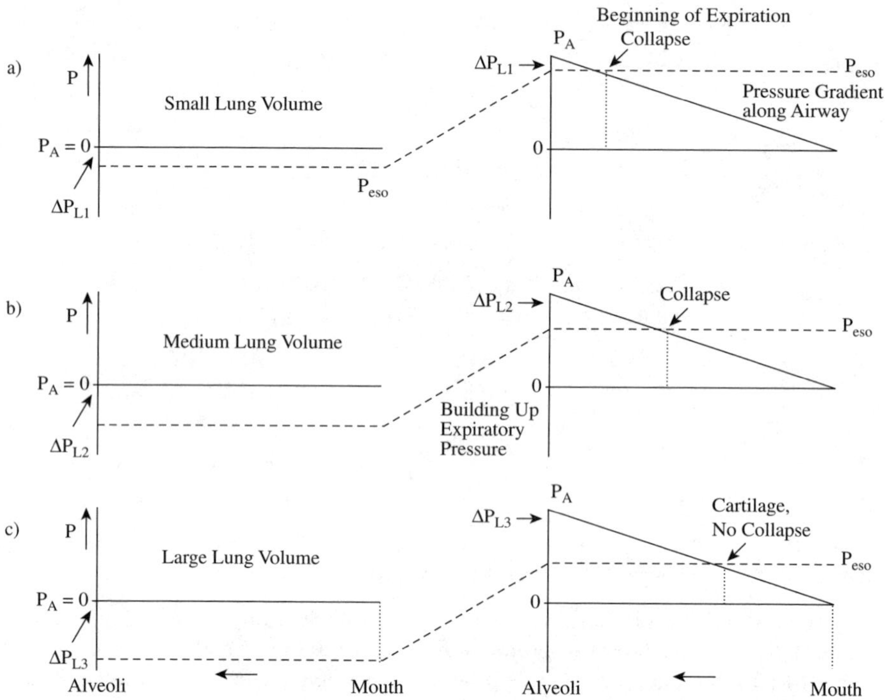

Figure 6.15. No-flow pressure gradients along the airways at the end of inspiration (a) and the start of expiration (b). Between these moments, the respiratory muscles build up intrathoracic pressure.

Airway collapse occurs where the intrathoracic pressure (measured as P_{eso}) is higher than the pressure inside the airways: deeper in the lung at lower volumes. At low lung volumes, this collapse arises in small airway passages close to the alveoli. At high lung volumes, however, no collapse occurs because the large airway passages close to the mouth are protected from collapse by their cartilageous rings.

Figure 6.15 shows the situation at one moment only. During the forced expiration, the outflow of gas causes the lung volume to decrease, which causes the point of collapse to move. Some parts of the lung may never be emptied by a forced exhalation; this is the "closing volume" (see Section 2.9).

Question: Patients with emphysema have airways that are very compliant and that collapse easily. Patients with asthma have airways that are constricted; collapse restricts the airways even more. Both have a tendency to breathe at large lung volumes. Why?

Airway collapse only occurs with *forced* expiration. With quiet normal breathing, the expiratory muscles are not used and no intrathoracic pressures exist that are higher than the pressures within the airways.

6.5. MEASURING DIFFUSION

The lung's O_2 diffusion capacity D_{O2} is defined as the O_2 flow through the lung membrane divided by the O_2 partial pressure difference across the alveolar membrane (see Section 1.2.8). The average O_2 flow through the lung membrane is the patient's O_2 consumption, which can be measured by spirometry. To establish the O_2 partial pressure difference across the lung membrane, we would need to measure the O_2 partial pressure in the alveolar gas and the O_2 partial pressure in the blood that perfuses the alveoli. The former can be determined from the end-expiratory O_2 concentration, the latter in principle from a blood sample. Since blood samples can be obtained from large blood vessels only, it would not be completely clear whether the arterial or the venous P_{O2} should be used or an average of both.

The following non-invasive method is most often used clinically. We know that carbon monoxide has almost the same diffusion properties as O_2 when it passes the alveolar membrane, i.e., $D_{CO} \approx D_{O2}$. We also know that CO, once in the blood, is rapidly bound to hemoglobin (210 times better than O_2); its partial pressure in the blood is therefore effectively zero. We therefore use CO instead of O_2, but since CO is toxic, very low concentrations must be used. The patient inspires a gas mixture which contains traces of CO (0.3%) and He (10%); the inspiration is from RV to TLC. After 10 s, the patient exhales again to RV. Because of He's very small diffusion capacity, we expect the inspired He to be completely exhaled again. We also expect the CO to have partially disappeared. During the 10 s, the alveolar CO concentration must have decreased exponentially (see Figure 6.16).

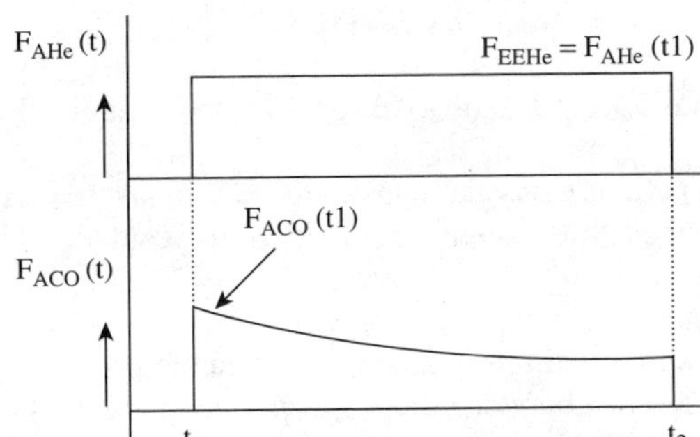

Figure 6.16. CO diffuses rapidly through the alveolar membrane, so that its alveolar concentration decreases. Helium cannot pass the lung membrane, so its alveolar concentration remains unchanged.

Question: Why does the alveolar CO concentration decrease exponentially?

The CO flow from lung to blood is

$$d/dt\ V_{CO}(t) = V_A\ d/dt\ F_{CO}(t) = D_{CO}\ F_{CO}(t)\ (P_{atm} - P_{H2O}) = D_{CO}\ P_{CO} \qquad (6.7)$$

where V_A is the alveolar volume, V_{CO} is the CO partial volume, F_{CO} is the CO fraction in the alveolar gas, and P_{CO} is the CO partial pressure. We also assume that the patient's glottis remains open, so that $P_{lung} = P_{atm}$. A correction for P_{H2O} has also been carried through. Formula 6.7 resembles Ohm's law, in which D_{CO} has the character of a resistance.

Integration from t_1 to t_2 (the test period) gives

$$F_{CO}(t_2) = F_{CO}(t_1) \exp(-D_{CO}(P_{atm} - P_{H2O}) / V_A)(t_2 - t_1) \qquad (6.8)$$

which we can solve for D_{CO}. $F_{CO}(t_2)$ is measured as the end-expiratory CO fraction, and V_A is computed from the inspired and expired He fractions (see Section 6.1.2.2). Still unknown is $F_{CO}(t_1)$, however. Its value follows from the known initial concentration ratio between CO and He

$$F_{CO}(t_1) / F_{ICO} = F_{He}(t_1) / F_{IHe} = F_{He}(t_2) / F_{IHe} \qquad (6.9)$$

where F_I stands for the known fractions in the inspired gas (0.3 and 10%).

Note that this method expresses the diffusion properties of the alveolar membrane as a single number. This number cannot give insight into any possibly existing local differences.

Many of the measurements that can be obtained by spirometry can also be performed when an unconscious, muscle relaxed patient is being ventilated. In a patient monitoring context, the inspired tidal volume and the inspired gas mixture can be manipulated, and the expiratory flow (and volume) and gas concentrations can be measured. Hence, many of the spirometry methods can (and have been) converted to practical instruments for monitoring operating room and intensive care unit patients.

QUESTIONS

1. Draw a schematic diagram of a spirometer and explain the function of each component.
2. Explain how a patient's oxygen consumption is measured with a spirometer.
3. The text describes three methods to measure residual volume. Name these methods. Explain one of them in detail.
4. Explain the shape of a typical capnogram.
5. Describe the single breath nitrogen washout method.
6. What is the difference between the static and the dynamic lung compliance?
7. Explain why the MEFV curve has an effort-independent part.
8. How is the lung's diffusion capacity measured?
9. Explain how a change in thorax compliance will likely change the lung's vital capacity.
10. Research possible modifications of the spirometry methods described in this chapter so that they could be converted to monitoring methods for patients who cannot cooperate, e.g., patients in an operating room or intensive care unit.

7 Circulatory Measurements

The problem of how well the circulatory system functions is mainly approached through measurements of blood flow, blood pressure and blood composition and their derivatives.

7.1. MEASURING CARDIAC OUTPUT

A moment-to-moment measurement of the blood *flow* produced by the heart would provide the best possible information about the heart's pump function. Alternatives are to measure *stroke volume*, the blood volume that the heart pumps per beat, or *cardiac output*, the blood volume that the heart pumps each minute. Both provide similar information, since the heart rate is easily measured.

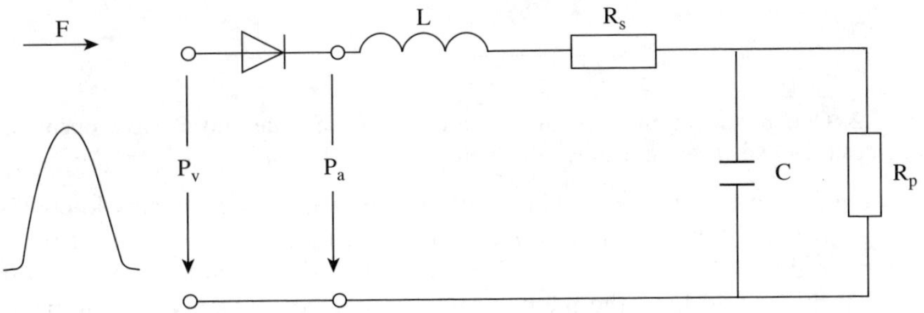

Figure 7.1. Model of the systemic circulation from left ventricle to vena cava.

One non-invasive method attempts to compute the cardiac output from the arterial pressure waveform, hence its name *pulse contour method*. It is based on the simple model of the systemic arterial circulation of Figure 7.1, which resembles a single wave rectifier circuit. P_v is the pressure generated by the left ventricle, the diode represents the aortic valve, and P_a is the pressure in the aorta. L is due to the inertance of the blood flowing into the aorta, C is the compliance of the total systemic arterial bed, R_s is the resistance that the blood inflow encounters and R_p is the systemic *peripheral resistance* due to the small blood vessels and capillaries that feed the peripheral tissues. All these parameters are variable; they depend on the pressures in the system and on the blood requirements of the organs. Thus, this model is only a coarse representation of reality. In a person at rest whose circulatory system is not too disturbed, the values of R_p and C will remain relatively constant. Cardiac output measurements are, however, most important in unstable patients where we cannot assume the constancy of these parameters.

Blood flows into the aorta only when the aortic valve is open. These points can be discovered in the arterial pressure waveform as the point where the systolic upstroke starts and the position of the dichrotic notch (Figure 7.2). Assuming that we perform a pressure measurement at a site that the model represents as the junction of C and R_p, the inflow F has two effects: it causes a flow through R_p and it increases the charge on C.

$$F = \frac{P}{R_p} + C\frac{dP}{dt} \tag{7.1}$$

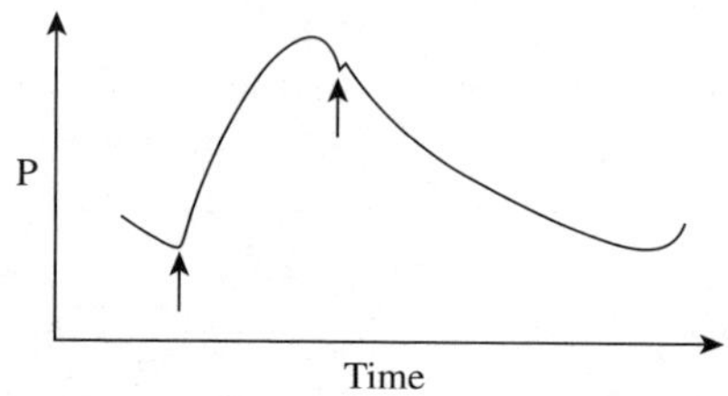

Figure 7.2. One period of the arterial pressure waveform. The aortic valve opens at the systolic upstroke and closes at the dichrotic notch.

Since we measure P and thus also know dP/dt, we can, if we somehow knew C and R_p, integrate F over the ejection period and obtain the stroke volume. In practice, however, we do not know C and R_p. We could use independently obtained average values, but this would limit the method to "average" persons only and eliminate its use for critically ill patients who are far from normal. We can also try to invent methods to somehow obtain the missing parameter values by simple methods. Note, for instance, that the model tells us that the diastolic pressure part of the curve, starting at the dichrotic notch, decreases exponentially. The time constant of this decrease provides us with the product of C and R_p, eliminating one unknown. Thus far, no reliable method has been discovered to eliminate the final unknown parameter.

This method is used primarily in those conditions where we can expect R_p to remain constant over long time periods. Since R_p is still unknown, the instrument is uncalibrated; its readout over time will show a *trend* that shows the variations in cardiac output, not cardiac output itself. Calibration, however, is possible if another method is also used occasionally.

7.1.1 Fick's method

Fick's principle, which is based on Fick's First Law (Section 1.2.8) states that all oxygen molecules in the lung venous blood originate either from the lung arterial blood or from oxygen transported from lung to blood. In Fick's method, the patient's oxygen consumption is measured, as well as the oxygen content[1] in the blood that enters the lung and in the blood that leaves the lung (Figure 7.3). This oxygen is, of course, carried by the blood flow. The oxygen that enters the blood is

[1] The oxygen content of the blood is primarily in the form of oxygen bound to hemoglobin.

reflected in the oxygen content difference between lung arterial and lung venous blood. The method depends on the fact that no oxygen is consumed by the tissues along the blood's path from lung artery to lung vein. If we average over a sufficiently long time period, we can write

$$F_{O2} = CO\ C_{vO2} - CO\ C_{aO2} \tag{7.2}$$

or

$$CO = F_{O2} / (C_{vO2} - C_{aO2}) \tag{7.3}$$

where F_{O2} is the oxygen flow from lung to blood, and C_{vO2} and C_{aO2} are the oxygen contents in lung vein and lung artery. F_{O2} can be determined by spirometry (see Section 6.1), while both C_{vO2} and C_{aO2} can be determined from blood sampled at the appropriate locations.

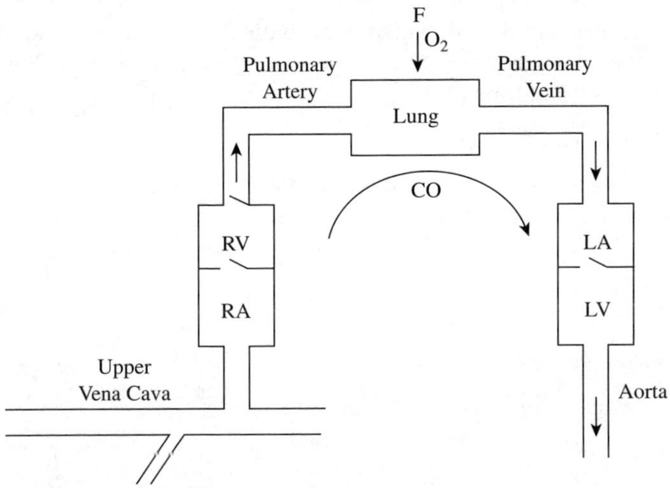

Figure 7.3. The oxygen content in the lung vein equals the oxygen content in the lung artery plus the oxygen that entered the blood from the lung.

Averaging over a sufficiently long time is necessary because the concentrations fluctuate due to the pulsatile nature of both respiration and circulation. The arterial oxygen content can be sampled at any convenient location in the arterial system (any large artery). Since oxygen transfer to the tissues does not occur in the arteries but in the capillaries, the blood's oxygen content is the same throughout the systemic arterial system. At the venous side, this is not true. The oxygen content in a vein depends on how much oxygen the organs, whose deoxygenated blood the vein transports, have extracted. Thus, the oxygen content may vary a great deal between veins. The *mixed venous* oxygen content is not available before the right atrium, in whose close proximity a number of veins merge. Mixed venous blood can therefore only be sampled in the right atrium, the right ventricle or preferably the lung artery itself, where the blood is well mixed by the right ventricular contraction. All these sites are accessible with a flotation (Swan-Ganz) catheter.

7.1.2. Indicator dilution methods

In indicator dilution methods, a bolus (brief injection) of an *indicator* (an easily and accurately measurable substance) is brought into the bloodstream, and the indicator's concentration is measured downstream. The path from the point of injection to the measurement site should have no branches through which some of the indicator could disappear or through which dilution could take place, and the tissues should not absorb it. Many different indicators are in use: chemicals, inert gases, radioactive isotopes, dyes (dye dilution) and heat (thermodilution). Indicators should not be toxic. They should not disappear (e.g., through metabolization into other products) during the duration of the test, but after the test they should disappear from the body — preferably in a natural way.

Assume that a small known bolus of dye is injected into the right atrium. Due to the heart's contractions, the dye is well mixed with the blood. Assume also that we measure the dye concentration in the aorta. The mass of dye that passes the measurement site during time dt is

$$dM = F(t)\,c(t)\,dt \tag{7.4}$$

where F is the blood flow and c the dye concentration. If one waits long enough, all dye will have passed the measurement site. Thus,

$$M = \int_0^\infty F(t)\,c(t)\,dt \tag{7.5}$$

It is only the average flow — the cardiac output — that determines how much dye is transported, not the fluctuations of the flow. We can therefore rewrite the above equation as

$$M = CO \int_0^\infty c(t)\,dt \tag{7.6}$$

Since M is known and c (t) can be measured and integrated, we can compute CO.

We will consider the dye transport in some more detail. To simplify the analysis, we assume that the flow is constant and that the flow profile is flat ("plug flow"). Even in this case, the molecules of the dye will not all have the same velocity. This is due to diffusion, which adds a random velocity component. The concentration can be described by a one-dimensional diffusion equation which is called *Fick's Second Law*, which is the time derivative of Fick's First Law. Fick's Second Law must be used when concentrations vary in time.

$$\partial C / \partial t = D\, \partial^2 C / \partial x^2 \tag{7.7}$$

Its solution, assuming an injection which initially causes a narrow normal (Gaussian) distribution, is a normal distribution

$$C(x,t) = \frac{\sigma_0}{\sigma} \exp - \frac{(x - vt)^2}{2\sigma^2} \tag{7.8}$$

with $\sigma^2 = \sigma_0^2 + 2\,D\,t$.

C (x, t) is the concentration at distance x from the injection site at time t, σ_0 the width of the initial distribution, σ the width of the distribution at time t, v the mean flow velocity and D the diffusion coefficient. Thus, a diffusing "cloud" of dye, moving at velocity v, has a normal distribution whose "width" (variance) increases as x (and t = x/v) increases. Figure 7.4 shows this.

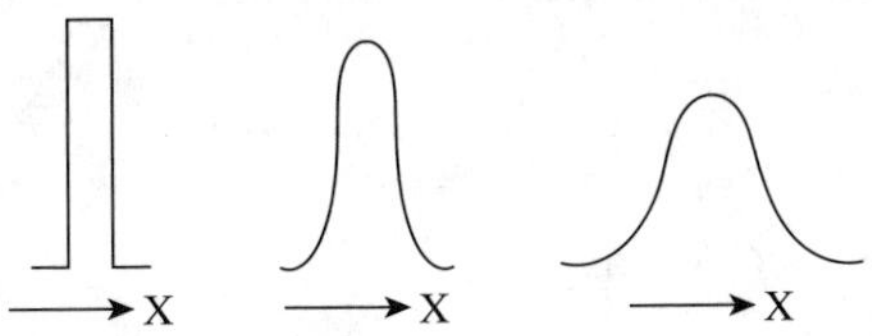

Figure 7.4. Due to diffusion, the "cloud" of injected dye broadens as time passes.

The curve of Figure 7.4 is not, however, the curve that is acquired in a measurement. The measurement is a function of t only, at a *fixed* x. Solving the above expression for a fixed x shows that the resulting curve is asymmetric. The reason is that the tail of the cloud passes at a later time, when the cloud is wider. Practice shows that, if mixing is adequate, the tail of the curve decreases, to a very good approximation, exponentially.

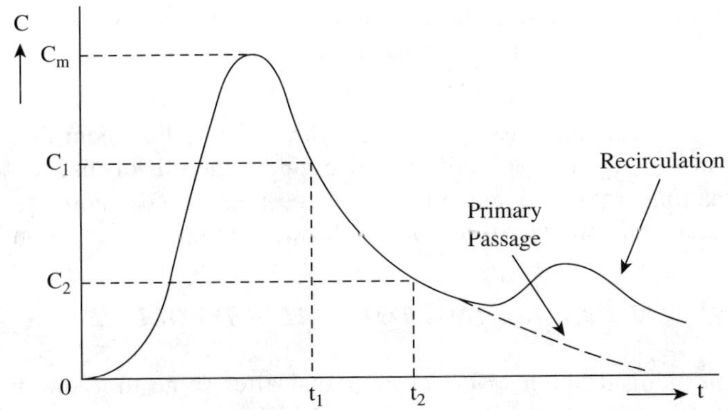

Figure 7.5. The concentration curve, measured at a fixed position, is asymmetric. If there is recirculation, it changes the tail of the curve.

Figure 7.5 shows the asymmetry of the dye concentration as it passes the sensor. It also shows another phenomenon: recirculation. While the tail of the cloud is still passing the measurement site, its onset has — along some shortest path — already arrived there for the second time. If recirculation exists, the integration from t = 0 to t = ∞ in formula 7.5 would obviously be incorrect. In that case one can use the exponential character of the final part of the curve. In Figure 7.5, the decrease between t_1 and t_2 is exponential; after t_2 recirculation starts. Up to t_2, the standard integration can take place. The integral of an exponential curve starting at t_2 would have as its value

$$I_{t2} = \int_{t_2}^{\infty} c\,(t)\,dt = \int_{t_2}^{\infty} C_2 \exp\left(-\alpha\,(t-t_2)\right) dt = C_2 / \alpha \tag{7.9}$$

where I_{t2} is the area under the curve starting at t_2 and α the curve's time constant. Both are unknown. Likewise, the integral starting at t_1 has a value $I_{t-1} = C_1 / \alpha$, and the integral starting at t_1 and ending at t_2 has a value $I_{t1-t2} = (C_1 - C_2) / \alpha$. The latter can, of course, be measured. This makes I_{t2} computable.

$$I_{t2} / I_{t1-t2} = C_2 / (C_1 - C_2) \tag{7.10}$$

or

$$I_{t2} = I_{t1-t2}\, C_2 / (C_1 - C_2) \tag{7.11}$$

In practice, once C_m has been found to be the maximum concentration, two points are chosen such that $C_1 = c_1 C_m$ and $C_2 = c_2 C_m$, with known constants c_1 and c_2.

In the thermodilution technique, a known quantity of "cold" (ice-cold water) is injected. This "cold" is carried along with the blood, and practice has shown that only very little of it is taken up by the vessel walls. A flotation catheter rapidly injects the cold water at the desired site, against the flow to ensure good mixing. At the catheter tip, at some 4 cm distance from the injection opening (larger distances cause too much "cold" loss), a thermistor (temperature-dependent resistor) measures the temperature. The temperature profile is processed as described above. Recirculation is non-existent. This technique makes blood flow measurements possible in any vessel that is accessible with a catheter.

A variant of this technique uses a catheter with a thermal element that produces a small heat "noise" signal (not a cold impulse) and a thermistor that senses the very small fluctuations in blood temperature. A correlation technique processes the measured signal and generates a continuous flow (and thus cardiac output) reading.

7.2. MEASURING TOTAL BLOOD VOLUME

Recirculation along different paths ensures that after some time the dye has been spread uniformly throughout all blood vessels. If the dye cannot escape through the vessel walls, a blood sample taken anywhere yields the total blood volume, using the formula $V = M / C$, where V is the total blood volume, M the mass of the injected dye, and C the dye concentration.

7.3. MEASURING THE BLOOD VOLUME OF ISOLATED VASCULAR BEDS

The dilution curve can also be used to determine the blood volume of an isolated vascular bed with one entry and one exit. One injects at the point of entry and measures at the exit point. Since there may be many different paths from entry to exit, each with its own flow and its own transit time (Figure 7.6), the curve that is measured will often not have the form of Figure 7.5. To compute the total blood flow F through a vascular bed, that is not a problem, because only the integral of the curve is required in its computation. But the form of the curve also contains information about the total blood *volume* V of the vascular bed. This volume $V = F\ t_m$, where t_m is the dye's mean transit time, i.e., that time that divides the total area under the curve into two equal areas (Figure 7.7).

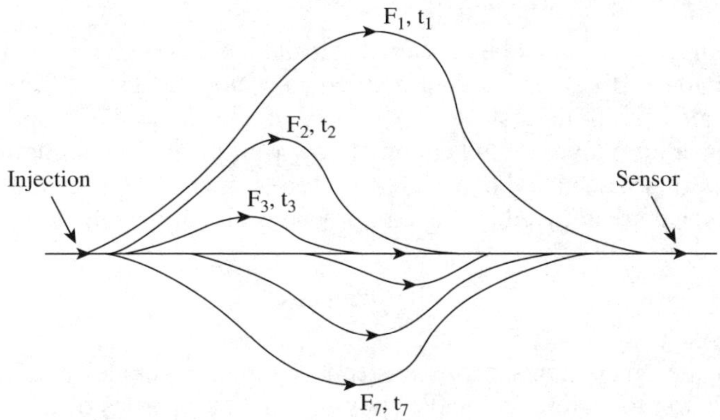

Figure 7.6. The vascular bed of an organ usually has one entry and one exit vessel, but many paths from entry to exit, each with its own flow and time delay.

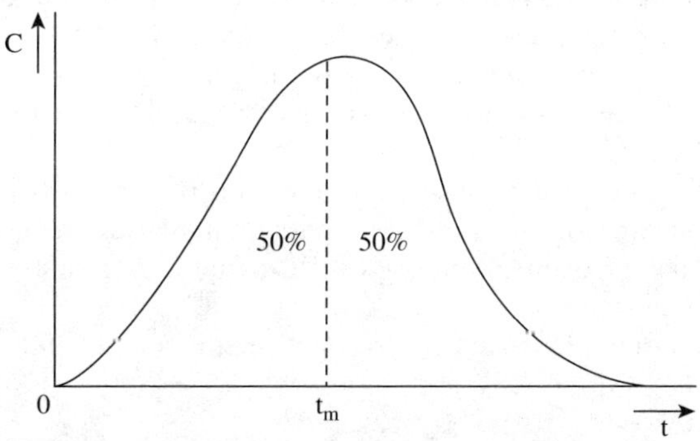

Figure 7.7. The mean transit time of the dye allows computation of the blood volume of an isolated vascular bed if the total blood inflow or outflow of the bed is known as well.

Significant rapid recirculation will make the calculations of F and V impossible or unreliable. The usual conditions apply: good mixing, and no loss of dye. The flows are also assumed to be constant during the test.

7.4. MEASURING TEMPERATURE

Temperature measurements are extremely simple and robust. The sensor usually consists of one or two simple thermistors (temperature-dependent resistors) or thermocouples (the junction of two different metals generates a temperature-dependent voltage). The term "core temperature" indicates the temperature deep inside the patient's tissues; it can be measured in the rectum, esophagus, nasopharynx or at the tympanic membrane; these give slightly different readings.

The core temperature needs to be monitored especially to detect undercooling in (badly thermally insulated) patients who undergo surgery for prolonged periods, since anesthesia inhibits the patient's temperature regulation. In rare cases, a patient may show "malignant hyperthermia" as an allergic reaction. This menacing condition must be detected as rapidly as possible.

The "skin temperature" must be measured at a site that is relatively isolated from the room temperature; if not, its readout may contain large artifacts (disturbances) due to the OR's air-conditioning system. A variety of sites is used. Comparison of core and skin temperature gives an indication of how well the peripheral tissues under the sensor are perfused. Skin above well-perfused tissues is warm; the temperature above tissues that are not well perfused approaches that of the OR.

QUESTIONS

1. Explain the components of the systemic circulation model of Figure 7.1 and their relation to anatomical and/or physiological properties of the human body. Describe at least three simplifications that were used to arrive at this (very simple) model.
2. The "pulse contour method" has been proposed as a method to track changes in cardiac output. Describe the model that is the foundation of this method and how the computation of the cardiac output is done using this model.
3. Describe Fick's method to measure cardiac output. Discuss the best locations to obtain blood samples.
4. Derive the formula that allows the calculation of the cardiac output from the observed concentration of an indicator passing a measurement site.
5. Derive formula 7.7 from formula 1.17. Research Fick's first and second laws if necessary.
6. Show why recirculation need not be a problem in dye dilution cardiac output measurements.
7. How can total blood volume be measured?

8 Therapeutic Devices

A large variety of therapeutic devices is available to clinicians. We will describe some that play a major role in the management of respiratory and circulatory problems.

8.1. SYRINGES, INFUSION DRIPS AND INFUSION PUMPS

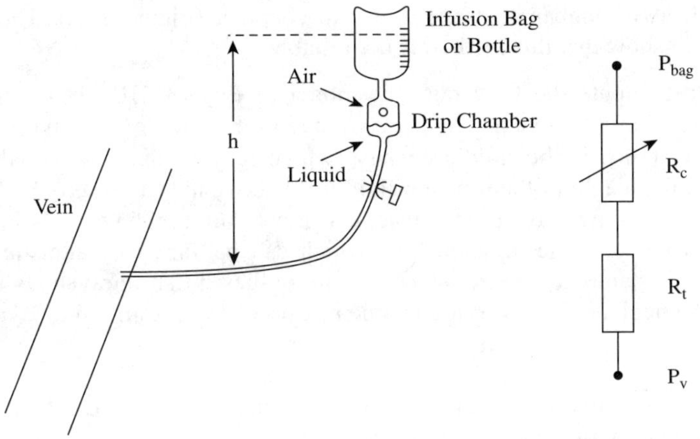

Figure 8.1. Infusion drip system (left) and its model (right). The drip chamber contains a volume of air, so that one can see individual drops of infusion fluid fall. If an infusion is delivered by a drip system, the infusion flow is often given as number of drops per minute.

Syringes are used to rapidly provide the patient with a known volume of one of a large variety of drugs. For fast action, drugs can be injected into the bloodstream, usually intravenously. For slower drug action, intramuscular injections can be given; their disadvantage is that the rate at which the drug becomes effective will depend on where the bolus was deposited in the tissues and on usually unknown characteristics of the patient, such as muscle bulk. Although an injection can directly influence a measurement (see, e.g., Section 3.2.1), their indirect influence — through the action of the drug — is much more important. Since many drugs influence the values of the measured variables, a correct interpretation of the measurements is possible only if it is known which drugs have been applied and in what dosage. Providing this information to a monitoring system is important for reliable record keeping, but especially when "intelligent" alarms are generated. In that case, the information must be provided to the system *at the time the injection is given*. Although bar codes on syringes and bar code readers are sometimes used, the requirement that no extra effort is to be demanded from the clinician makes this approach unpopular and error-prone.

Infusion drips realize a slow, yet carefully controlled drug flow. A bag or bottle, usually filled with a physiological salt solution to which a known volume of the drug is added, empties itself slowly through a catheter into the patient, often into a vein. The desired flow rate is adjusted through a restriction in the catheter and can be read off as a number of drops in a drip chamber (Figure 8.1a). The following analysis will show that this method is not reliable.

In order to compute the flow rate, the model of Figure 8.1b can be used, which assumes infusion into a vein. The pressure due to the bag P_{bag} is a hydrostatic pressure ρgh, where ρ is the fluid's density (which may be assumed equal to that of water), h is the height of the infusion fluid level above the site where the fluid enters the vein, and g is the gravitation constant. The pressure in the vessel is P_v. R_c stands for the resistance of the infusion line, which can be varied by adjusting the flow restriction. R_t stands for the resistance of the tissues; in an intravenous infusion R_t can often be neglected. By varying R_c, the infusion flow is adjusted to a value

$$F = (P_{bag} - P_v) / (R_c + R_t) \qquad (8.1)$$

We see, however, that F will change if P_v or R_t changes. If this can be the case, the flow rate cannot be accurately maintained.

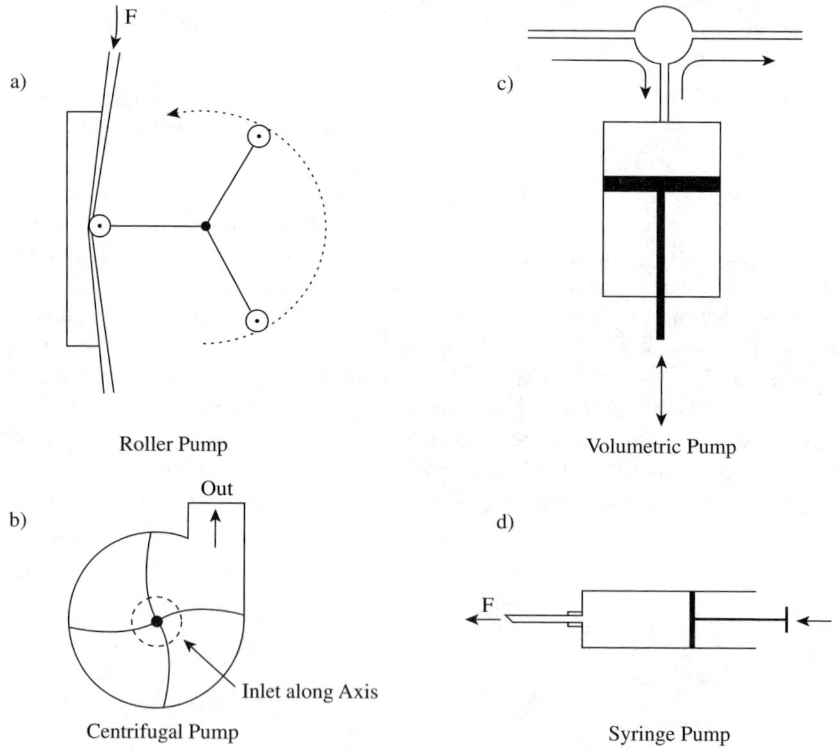

Figure 8.2. Roller pump (a), centrifugal pump (b), volumetric pump (c), syringe pump (d).

An infusion flow *into* the patients requires that P_{bag} is larger than the pressure inside the vessel P_v. A lower value of P_{bag} would cause backflow of blood into the infusion line. Infusion into a *vein* can be realized by suspending the infusion bag at a certain height above the vessel. Infusion into an *artery* cannot be done this way. Since an arterial pressure of 200 mm Hg corresponds to a water column of 272 cm high, the height of the bag would have to be almost 3 m above the blood vessel. Few rooms have that height. In arterial infusions, the infusion bag is therefore enclosed in an extra inflatable bag, which is pressurized to 200 or 300 mm Hg. The infusion flow rate depends on fluctuations of the arterial pressure. In arterial infusions, a high pressure difference $P_{bag} - P_v$ may exist; an extremely reliable, constant value of the flow restriction R_c must prevent accidental infusions of too large flows.

Infusion pumps can realize accurately known flow rates. Figure 8.2 shows some types. In a roller pump, one of its wheels moves the point of collapse in a flexible tube, thus pushing the fluid in the tube forward. In a centrifugal pump, the spin of the rotor creates a centrifugal force that drives the fluid toward the walls and the outlet; fluid enters along the axis to replace the fluid that is pumped out. A volumetric pump has two cycles. In a rapid cycle, the pump's plunger downward movement fills the pump's chamber with infusion fluid from the bag. In a slow cycle, the plunger moves upward and the chamber's fluid is infused into the patient. Unidirectional valves allow flow in one direction only. In a syringe pump, a (large) syringe is slowly emptied into the patient.

All these pumps can be modeled as flow sources, where the flow that they generate is independent of resistances and counterpressures. High pressures arise when the line is partially or fully occluded. The pressure in the line is therefore often measured at the infusion pump; it can be used to generate an alarm and/or to switch off the pump if the pressure becomes too high.

8.2. VENTILATORS

When patients undergo major surgery in the OR, their muscles must be relaxed. As a consequence, the respiratory muscles are relaxed as well, and the patient is not capable of breathing. Artificial respiration by a ventilator or anesthesia machine[1] is then required. Ventilation is also required in ICU patients whose respiratory musculature is too weak to provide adequate spontaneous ventilation. Basically, during inspiration a pressure or flow source forces gas into the lungs through the *endotracheal tube*, which is inserted into the patient's trachea. An inflatable cuff around the end of the endotracheal tube forms an airtight connection. During expiration, the pressure or flow source is removed and the lung will passively deflate, due to the compliance of lung and thorax.

A variety of ventilator types exist. We will discuss the two most popular types. Most ventilators can provide either an airway *flow* or a positive airway *pressure* to the

[1] Sometimes the term *ventilator* is used for that part of an anesthesia machine that generates the flow. We will use the terms "ventilator" and "anesthesia machine" interchangeably.

patient. The difference becomes clear when we view the simplest possible model for the inspiratory phase of a ventilator (Figure 8.3). R represents the patient's airway resistance, C the combined lung-thorax compliance (see Figure 1.9). During a fixed time interval T_{insp} (the inspiration time), the ventilator either provides a constant flow F, resulting in a pressure P, or it provides a constant pressure P, resulting in a flow F.

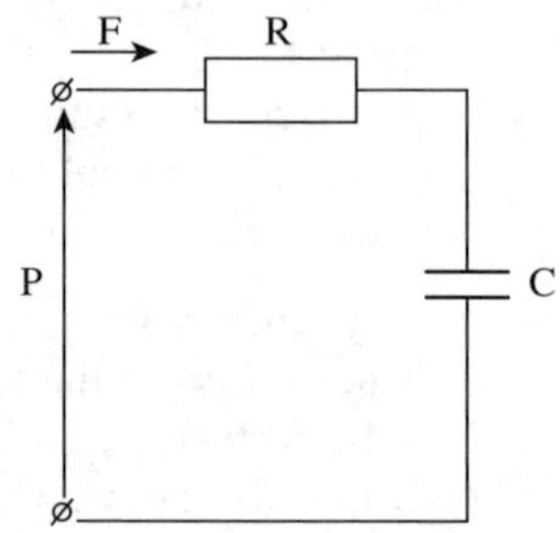

Figure 8.3. A ventilator generates either a pressure P or a flow F. In either case, the other variable will depend on the patient's airway resistance R and lung-thorax compliance C.

If the ventilator provides a constant flow F during time T_{insp}, the patient's tidal volume TV is guaranteed to be constant at the value F T_{insp}, regardless of the values of R and C. The pressure P, however, depends on these values. Its maximum value P_{max} will occur at the end of inspiration and have the value

$$P_{max} = F\ R + TV / C \tag{8.2}$$

A high value of P_{max}, which occurs when R is high and/or C is low, can cause barotrauma and is to be avoided.

If the ventilator provides a constant pressure P during time T_{insp}, barotrauma is avoided. The tidal volume TV, however, critically depends on the values of R and C.

Question: Express TV as a function of R, C, P and T_{insp}.

The consequence of constant pressure ventilation may be that insufficient oxygen is delivered to the patient. Even if the tidal volume is adjusted to the correct value initially, problems may arise when R or C change over time. Inadequate oxygen supply can be corrected by changing the inspired oxygen concentration.

In practice, both these problems can be avoided by accurate monitoring, because ventilators routinely provide the numerical values of both maximum airway pressure and tidal volume. Flow-controlled ventilation seems to be most frequently used, however.

8.2.1. The inlet combination

The patient must be provided with the appropriate gas mixture. In intensive care units, the inspired gas is usually a 30% O_2-containing mixture of air and pure oxygen, both of which are available from wall plugs fed by the hospital's gas supply system. The air is enriched with extra oxygen as a margin of safety to compensate for possible respiratory problems. Up to 100% O_2 can of course be given, but doing this for long periods will harm the lung tissue. In operating rooms, the inspired gas is usually a 30% O_2-containing mixture of pure oxygen and nitrous oxide (N_2O), an anesthetic, also available from wall plugs. In case of failure of one or more gases of the hospital's gas supply system, built-in gas cylinders are used as backup.

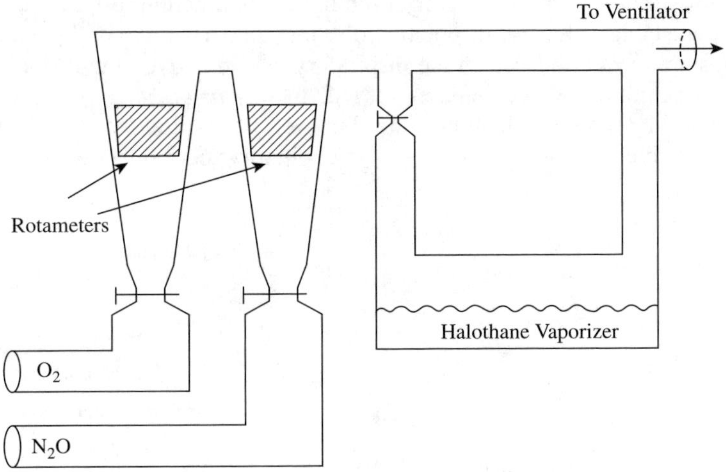

Figure 8.4. The inlet combination of an anesthesia ventilator. The gas flow to the patient is a combination of adjustable flows of oxygen, nitrous oxide and anesthetic agent.

The flow of both individual gases is regulated by proportional valves and measured by rotameters (see Section 4.1.3) in the so-called *rotameter block*. Since even 70% N_2O is insufficient to anesthetize patients, the combined gas (or part of it) is led through a vaporizer, where some anesthetic agent (e.g., 0 to 2% halothane) can be added to the gas (Figure 8.4). Sometimes water vapor is also added to the gas by a similar vaporizer (humidifier). Note that the flow of gas that leaves the so-called "inlet combination" is *constant*.

8.2.2. Open systems

Figure 8.5 shows a so-called "open system," where open refers to the fact that the exhaust of the system is open; expired gases are simply disposed of (not into the air of the operating room, due to the N_2O, but into a scavenging system). An *endotracheal tube* is inserted into the patient's trachea and a rubber balloon at its end is inflated to provide an airtight connection. The Y-piece connects the endotracheal tube with the inspiratory and expiratory hoses, which provide a flexible connection to the ventilator.

The ventilator has two valves, the inspiratory valve and the expiratory valve, whose opening and closing times are determined by an electronic control unit. During expiration, the expiratory valve is open, and the gas in the lung can escape through the expiratory tube into the scavenging system. During expiration, the inspiratory valve is closed, and the fresh gas from the inlet combination is collected in the bellows, which fills up. Because a spring attempts to keep the bellows at zero volume, the pressure inside the bellows rises. When the inspiration starts, the inspiratory valve opens and the expiratory valve closes simultaneously. Both the fresh gas from the inlet combination and the fresh gas that was collected in the bellows during the previous expiration now have a chance to flow into the patient's lung. A flow limiter in the inspiratory limb has as its function to limit the flow to at most an adjustable value F_I; in practice, the inspiratory flow has this value during the full inspiratory period. When the inspiratory period is over, the inspiratory valve closes; the expiratory valve remains closed. In this *inspiratory pause* period, the lungs remain insufflated. When this period is over, the expiratory valve opens — the inspiratory valve remains closed — the expiratory period starts and the cycle repeats.

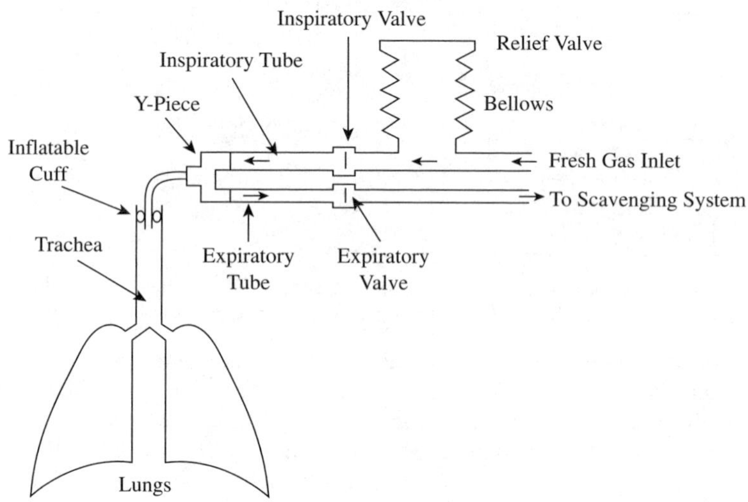

Figure 8.5. In an open system, the expiratory gas is not reused.

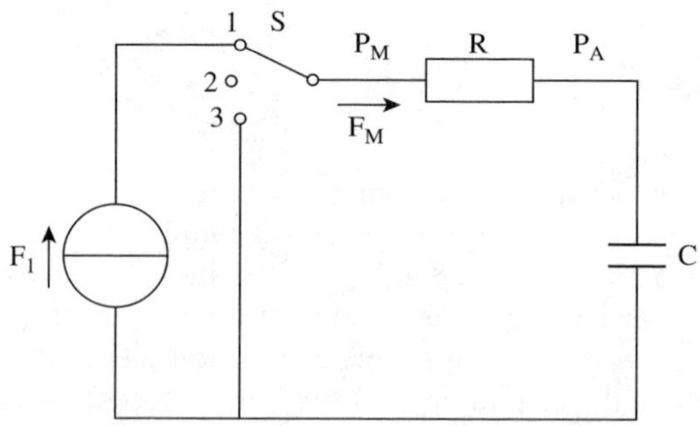

Figure 8.6. Model of an open system. Position 1 of switch S corresponds with inspiration, position 2 with the inspiratory pause, and position 3 with expiration.

If we disregard the resistances of the ventilator's tubes, we can model the operation of this open system with the diagram of Figure 8.6. In the upper position of switch

S, the inspiratory flow F_I charges capacitor C through resistance R (the inspiratory period). In the middle position, there is no flow (the inspiratory pause period). In the lower position, capacitor C can discharge through resistor R (the expiratory period). P_M is the pressure that is measured close to the patient's mouth, where the flow F_M is also measured. P_A is the alveolar pressure. During the inspiration, F_M is equal to F_I; during the expiration, it is determined by the way in which C discharges through R.

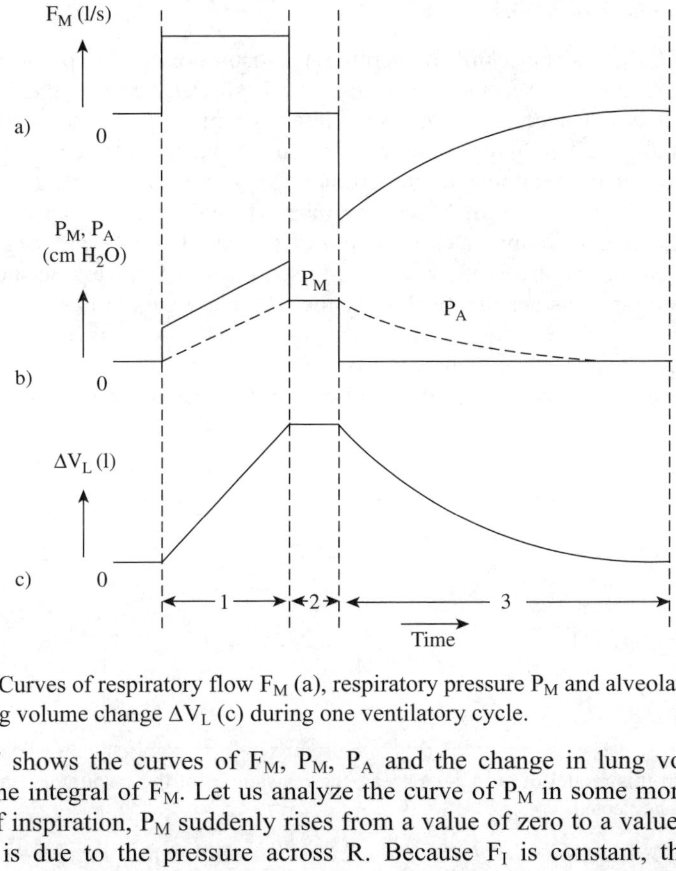

Figure 8.7. Curves of respiratory flow F_M (a), respiratory pressure P_M and alveolar pressure P_A (b) and lung volume change ΔV_L (c) during one ventilatory cycle.

Figure 8.7 shows the curves of F_M, P_M, P_A and the change in lung volume ΔV_L, which is the integral of F_M. Let us analyze the curve of P_M in some more detail. At the start of inspiration, P_M suddenly rises from a value of zero to a value equal to F_I R, which is due to the pressure across R. Because F_I is constant, this pressure remains constant throughout the inspiration. As the lung fills up, the pressure P_A due to its compliance C rises; because $P_M = P_A + F_I R$, this is visible in the slope of P_M during inspiration. During the inspiratory pause, the pressure drop due to R becomes zero, and $P_M = P_A$. A zero flow thus allows us to measure the alveolar pressure (in this case its maximum) at the mouth. During expiration, P_M is of course zero; P_A decreases exponentially.

During inspiration, the flow F_M is fixed by the ventilator setting, and during the inspiratory pause period it is zero. During expiration, F_M decreases exponentially; in the model, it is the current that is caused by the discharge of C through R.

Question: Assume that the model of Figure 8.6 is sufficiently accurate and that the measurements of F_M and P_M of Figure 8.7 are available. How can we reconstruct the values of the patient's R and C from the signals?

Question: It is consistently found that when the airway resistance R is computed separately during the inspiration period and the expiration period, the "expiratory resistance" is larger than the "inspiratory resistance." Can you think of a reason why this is to be expected?

In Figure 8.6 the resistances of the ventilator's inspiratory and expiratory tubes and hoses were neglected because these are usually much smaller than the airway resistance. It is not necessary to do so. Figure 8.8 gives a slightly more complete model, and Figure 8.9 gives a complex master-slave model (see Section 2.11.3) consisting of many sections. In the master (lung mechanics) model, resistances model the flow resistances of hoses and tubes (L and C terms can be neglected), capacitances model compliances of volumes (R and L terms are neglected) and (voltage or current controlled) switches are used to model valves. Because all slave (gas transport) models are identical, only the CO_2 slave model is shown. The slave model consists of current sources and capacitances only. A capacitance now models the total gas volume of a section; if the volume changes, a variable capacitance must be used. Current sources describe the transport of CO_2 from one section to the next.

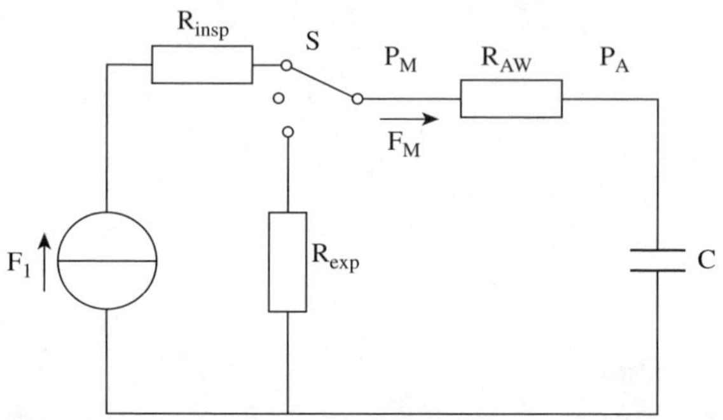

Figure 8.8. In this model of an open system, the resistances of the inspiratory and expiratory tubes are not neglected.

Question: If we use the model of Figure 8.8 rather than that of Figure 8.6, the measurements of F_M and P_M will differ from the curves of Figure 8.7. What are the differences?

Figure 8.10 shows that the modeled pressure and flow curves (Figure 8.7) of even a simple model (Figure 8.6) are good representations of reality. The smoothness of the curves of Figure 8.10 is due to sensor bandwidth limitations; low pass filtering of the model signals would make the similarity even greater. The Figure shows the ventilator in a mode called *assisted ventilation*. This mode is used when it is desirable that the patient breathe spontaneously, even though the respiratory muscles are too weak to provide adequate pressures. This is the case, for instance, after surgery, when muscle relaxation wears off. In such cases, the ventilator's minute volume is chosen somewhat — but not dangerously — small. The ensuing slight lack of oxygen (hypoxia) and slight surplus of carbon dioxide (hypercarbia) stimulate the patient to breathe spontaneously. Observations show that the patient attempts to increase the respiration frequency: the patient starts to inspire before the start of the ventilator's new inspiratory cycle. This inspiration causes an underpressure that can be measured at the Y-piece or in the ventilator, and it can be used to start a new inspiratory cycle immediately.

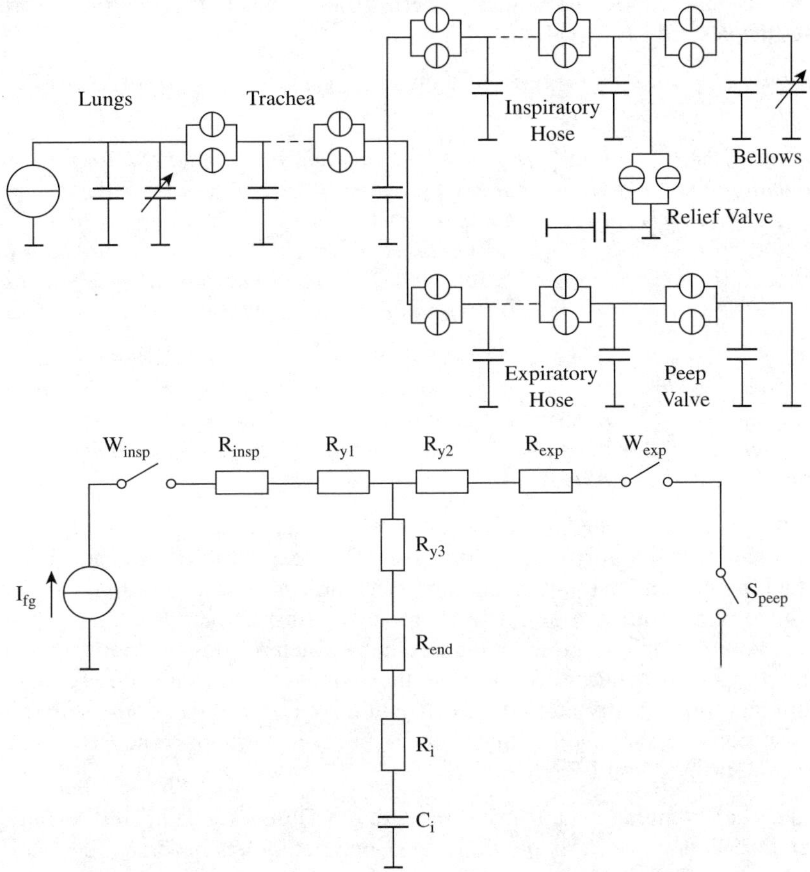

Figure 8.9. Multiple respiratory model: a pressure-flow model (slightly more refined than the model of Figure 8.8) and a gas transport model (see Section 2.11.2).

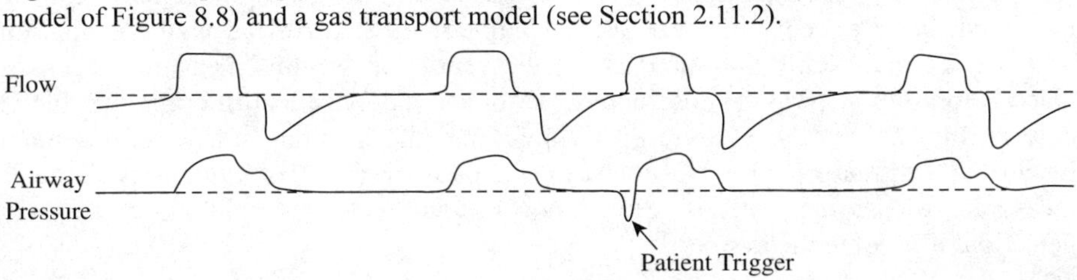

Figure 8.10. Registration of respiratory flow and pressure of a patient. The ventilator is in assist (or demand) mode.

Figure 8.7 shows that the alveolar pressure decreases to zero at the end of expiration. If lung collapse threatens, as in pneumothorax, this is not desirable. An extra ventilator option is a valve in the expiratory tube, which allows gas to pass only when the pressure is above some adjustable value (0 to 15 cm H_2O). This valve, which realizes a *positive end-expiratory pressure* (PEEP), is called a PEEP-valve. Another way to achieve PEEP is to make use of an *expiratory pause* — in a "pause" both valves are closed — which halts expiration when the expiratory pressure reaches the desired PEEP-value.

Question: How will the flow, pressure and lung volume curves of Figure 8.7 change when the PEEP is 10 cm H_2O?

Question: If the expiration time is not long enough, the patient cannot fully exhale. The remaining volume of gas in the lung causes an alveolar end-expiratory pressure which is called inherent PEEP. Compute this inherent PEEP when the alveolar end-inspiratory pressure is 10 cm H_2O, the expiration time is 4 s, the lung-thorax compliance $C = 0.15$ l/cm H_2O (approximately normal) and the airway resistance $R = 20$ cm H_2O s l^{-1} (approximately 10 times normal).

Question: Pneumothorax can exist in several forms. Assuming that the perforation is small, compare how severe lung collapse will be in both spontaneous respiration and artificial ventilation in two situations: a) the inner pleura blade is punctured; b) the outer pleura blade is punctured. Note: The (normally fluid-filled) space between inner and outer pleura blades is the intrapleural space.

Ventilation can be adjusted to the patient's requirements through a variety of means. The settings of inspiratory, inspiratory pause, and expiratory times can be adjusted to obtain the desired respiratory rate and I/E-ratio (ratio of inspiration + inspiratory pause time to expiration time). The choice of the inspiratory flow then determines the tidal volume and the minute volume. The rotameters allow the adjustment of the inspiratory gas concentrations and thus the ratio of O_2 to N_2O (OR) or O_2 to air (ICU); both are normally around ½, or about 30% O_2. In the OR, the setting of the vaporizer controls the inspired anesthetic agent concentration. The PEEP-valve, of course, determines the PEEP-pressure.

Adequacy of ventilation must be monitored. Observation of the expired gas concentrations is part of this monitoring. Capnography has become a standard. The end-expiratory CO_2 concentration or partial pressure (P_{ACO2}) is normally a good reflection of the P_{aCO2} (Figure 1.14). Oxygen and other gases and vapors can be measured, but frequently are not. Pulse oximetry is standard as well. Inspiration pressure is usually only measured inside the ventilator, but this measurement takes place before the inspiratory hose and does not accurately reflect the pressure that is delivered to the airways (Figure 8.8). Sometimes the inspiratory flow is measured inside the ventilator; more frequently the expiratory (tidal or minute) volume is measured. Measuring both is almost never done; yet, it would allow an easy detection of leaks in the system.

Hidden to the user but essential for safe operation are a number of safety features. Antibacterial filters remove bacteria and contaminations from the inspiratory gases. Overpressure valves at several sites open when pressures become too high. If the external gas supply fails, built-in gas cylinders provide the gases. In case of a power failure, an underpressure valve in the expiratory limb can allow the spontaneous inspiration of the patient to supply outside air, but only, of course, if the patient is capable of spontaneous respiration. If not (in the OR, where the muscles are relaxed), periodic squeezing of a self-inflating rubber bag by the anesthetist takes over gas delivery.

8.2.3. Circle systems

A disadvantage of an open system is that the expiratory gas is simply disposed of. It may, however, contain significant amounts of anesthetic agents, which are very expensive. In a "closed system" or "circle system," the expiratory gas is reused after its carbon dioxide has been removed (see Section 6.1) and its oxygen replenished. Advantages are that the inspiratory gas is water vapor saturated, which prevents dehydration during long operations, that the patient's cooling is reduced because warmer gas is inspired, that the environment is less polluted, and that fresh gas consumption is less than 5 l/min in a circle system, compared to more than 15 l/min in an open system.

Figure 8.11 shows a circle system, which also differs from the open system that we described in other respects. Whereas in the open system the valves were active (electronically controlled) and the bellows passive (uncontrolled), now the bellows is the active element — on inspiration, it is compressed by an outside pressure — and the valves are passive. The valves are mechanical devices (Figure 8.12) that allow gas passage in only one direction. In addition, a circle system needs a carbon dioxide absorber.

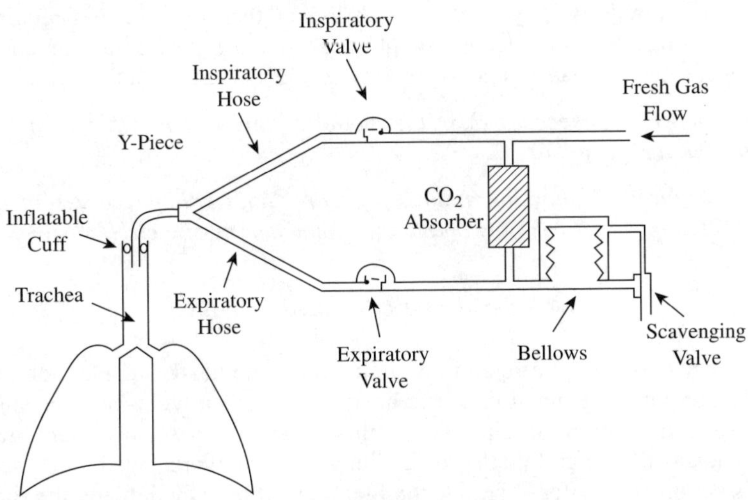

Figure 8.11. In a circle system, the expiratory gas is (partly) reused.

During inspiration, the bellows is forced downward by a pressure applied to its outside. This forces the gas inside the bellows through the carbon dioxide absorber

— the expiratory valve is forced closed — and through the inspiratory valve, which is forced open, into the patient's lung. During expiration, the pressure on the outside of the bellows is removed, and the lung empties itself passively through the expiratory hose, the expiratory valve — now the inspiratory valve is forced closed — and into the bellows. In this case, there is no inspiratory pause; whether one is possible depends on the ventilator. The fresh gas flow is the source which replenishes the oxygen.

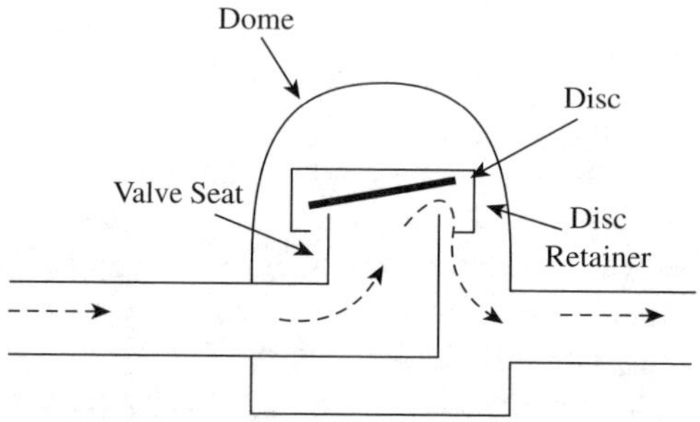

Figure 8.12. Passive inspiratory or expiratory valve. The mechanical construction allows gas to pass in one direction only.

In a fully closed system, only the patient's oxygen consumption (about 250 ml/min), and in the OR some anesthetic gas, would need to be added to the system. This would be another advantage of a fully closed system: the patient's oxygen consumption, an important metabolic indicator, would become known. Fully closed systems, however, are very difficult to realize; small leaks in the connections between tubes and hoses are almost impossible to avoid. With extreme care, fresh gas flow rates of 500 ml/min are possible; a value of 1 l/min is more common. Thus, the fresh gas flow is usually chosen much higher than the patient's requirements. Since overfilling the system leads to high pressures, the excess gas valve ("pop-off valve") opens and lets gas escape to the scavenging system when the bellows is full.

Question: In which respect(s) does the model of this system differ from the open system model of Figure 8.6?

Question: Figure 8.7 shows pressure, flow and lung volume curves of an open system. Sketch the pressure, flow and lung volume curves of a circle system.

8.3. HEART-LUNG MACHINES

Heart surgery is often delicate and possible only if the heart is made motionless. In that case, the pumping function of the heart is lost and a heart-lung machine must ensure the continuation of circulation; this is called *extracorporeal circulation* (ECC). The gas exchange function of the lung is taken over as well, because technically it is simpler and safer to realize the functions of both the left and the right heart with one pump. This also eliminates the thorax movements that are normally caused by ventilation. Figure 8.13 shows a diagram of a heart-lung machine.

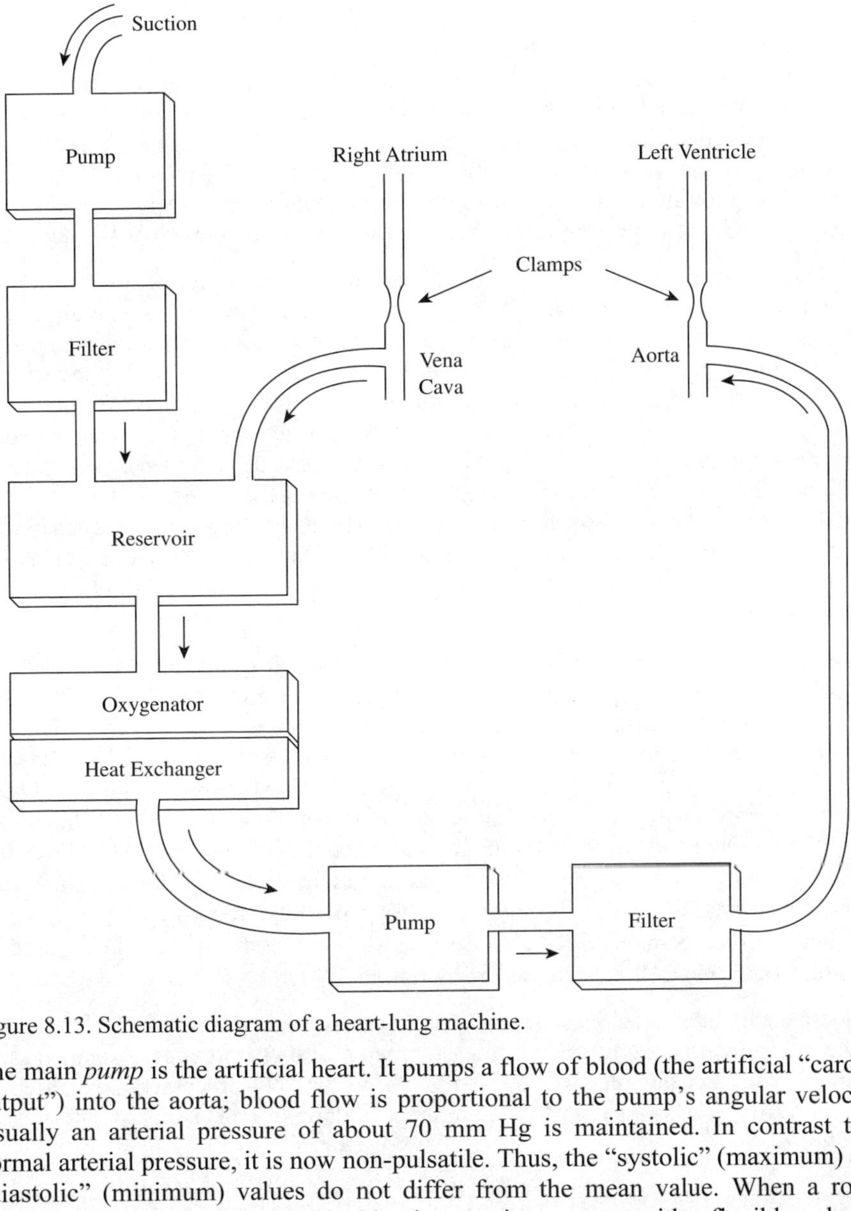

Figure 8.13. Schematic diagram of a heart-lung machine.

The main *pump* is the artificial heart. It pumps a flow of blood (the artificial "cardiac output") into the aorta; blood flow is proportional to the pump's angular velocity. Usually an arterial pressure of about 70 mm Hg is maintained. In contrast to a normal arterial pressure, it is now non-pulsatile. Thus, the "systolic" (maximum) and "diastolic" (minimum) values do not differ from the mean value. When a roller pump (see Section 8.1) is used, the blood comes into contact with a flexible nylon or polyester hose only. These materials limit damage to the blood as much as possible. Centrifugal pumps result in even less blood damage and can thus be used for longer time periods.

The *oxygenator* (gas exchanger) is the artificial lung. Normally, a membrane oxygenator is used. Blood flows on one side of the membrane; on the other side, a liquid circulates whose O_2 and CO_2 partial pressures are controlled in such a way that desired blood gas values result. This is done by bubbling adjustable flows of O_2, CO_2 and air through the liquid. Usually these flows are adjusted by and read off from a *rotameter block*. The advantages of membrane oxygenators are the avoidance of gas emboli (these are formed in a bubble oxygenator in which gas is bubbled through the blood) and minimal damage to the blood, of which especially the platelets (thrombocytes) are easily disrupted. Disruption of platelets, which causes coagulation (clotting) of the blood, cannot be prevented completely. Coagulation is attacked by adding large amounts of heparin (an anticoagulation agent) to the blood.

Additional pumps force leakage blood and lung venous blood that may have accumulated back into the circulating volume. Filters remove blood clots and air bubbles from the circulation. Flexible hoses connect the parts of the heart-lung machine together and thick cannulas connect the hoses to the patient's (inferior and superior) vena cava and aorta. The venous outflow is passive; it is due to the hydrostatic pressure difference that results from the lower position of the heart-lung machine's blood reservoir. A heat exchanger controls the blood temperature. The artificial heart forces the blood through the oxygenator and, after gas bubbles and blood clots have been removed, into the aorta. The heart-lung machine adds about 2 l to the circulating blood volume in an adult, half of this in children and babies. This extra volume is (donor) blood, blood plasma or plasma similar fluids (colloidal electrolytes).

Hypothermia (a reduction of the patient's temperature) to about 28°C reduces the tissue's oxygen consumption and carbon dioxide production with about 50%. This provides a safety margin in this artificial situation, where perfusion is far from normal. The heat exchanger consists of a heat exchange surface, usually of a metal coated with nylon (polyamid) or a similar material. Blood flows on one side. On the other side flows water, of which the temperature can be controlled. At the start of the operation, cold water cools the patient; at its end, warm water reheats the patient to 34°C. Most operations can be done at moderate hypothermia (down to 25°C). In some operations, the blood flow must be stopped completely. Since metabolism is reduced to about 50% at 27°C, about 6 min at most are then available for surgery. Deep hypothermia at 19°C extends this period to 30 min.

Modern heart-lung machines have a computer interface through which they can make their internal settings and measurements (flows, pressures, temperatures) available to the outside world. These data can be processed and used to automatically generate reports, "intelligent" alarms, and to automate certain functions.

8.4. DEFIBRILLATORS AND CARDIOVERTERS

The cause of mechanical heart failure, i.e., loss of pump function, is usually due to ventricular fibrillation. In this condition, the depolarization of the individual heart muscle fibers has become desynchronized. Resynchronization can be accomplished by sending an electric current through the heart muscle. This depolarizes all muscle fibers simultaneously, so that the normal rhythm will usually be reinstated. The device that generates the electric current is called a *defibrillator*. In the simplest type

(Figure 8.14), a capacitor, which has been charged to 500 to 800 V, is discharged through electrodes which are forcefully pressed to the patient's chest, above the heart. Switches in both electrodes ("paddles") must be depressed to deliver a pulse. Usually, one discharge is sufficient. A readout on the defibrillator indicates the energy $E = \frac{1}{2} C V^2$ that will be delivered to the patient. A large electrode surface area, a well-conducting contact gel and a forceful contact result in a chest resistance of only 50 Ω. The current through the chest can therefore have a value up to 20 A. Usually, a force sensor integrated with the electrodes measures the applied force and prevents a discharge when the pressure is too small.

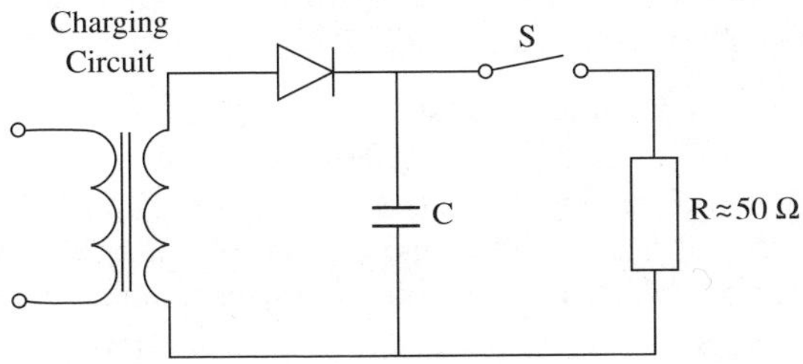

Figure 8.14. Schematic diagram of a simple defibrillator.

Experience with these devices has shown that too high and too low voltages are not effective. Too high voltages and currents may cause burns, due to heat generated by the contact resistance of the electrodes. The best pulse form to be delivered to the patient is a block pulse; its best duration is about 20 ms. The simple defibrillator of Figure 8.14 is thus far from ideal. Better designs incorporate one or more inductors, which decrease the pulse's maximum amplitude, or electronics.

In atrial fibrillation, the pump function of the heart is not lost; it is only less efficient. Defibrillation is now possible as well, but care must be taken that the defibrillator's pulse does not cause ventricular fibrillation, as any large current through the heart can. Therefore the pulse is delivered at a time when the ventricle is depolarized anyway. This moment is recognizable in the ECG by the R-wave. A defibrillator, which detects R-waves and delays the application of its pulse until an R-wave is detected in the ECG, is called a *cardioverter*, and this type of defibrillation is called *cardioversion*. The electrodes are now also used to acquire the ECG, and a built-in ECG monitor extracts the moments when R-waves occur.

In internal defibrillation (when the heart itself is accessible to the electrodes), a smaller voltage and smaller electrodes are used. Internal defibrillation is often required at the end of heart surgery, when the heart has returned to its normal temperature and can resume its normal function. In some cases, it restarts by itself.

Instruments that are attached to the patient must be able to withstand the high voltages and currents of defibrillators and cardioverters.

Defibrillators, or rather cardioverters, can also be implanted. An *implantable cardioverter defibrillator* (ICD) detects the presence of ventricular arrhythmias and fibrillation, and applies an electrical pulse to the ventricle when this is needed.

8.5. PACEMAKERS

The cause of electrical heart failure is often due to problems in the generation or conduction of depolarization in the heart muscle. The AV-node in particular is vulnerable. It consists of a bundle of very small nerve fibers. It is located between atrium and ventricle, and its function is to delay the depolarization and contraction of the ventricle until it has been adequately filled by the atrium. If its function is lost, the ventricle still contracts but at a far too low rate and with a very low cardiac output. This condition is called *heart block*. An implanted pacemaker can resolve this problem in patients with heart block. Before the implant operation, an external pacemaker can keep the patient in a fit condition.

A pacemaker can be programmed (in an implanted one non-invasively, through a "magnet" on the skin) into a variety of "pacing modes." In the simplest of these, a small electrical pulse (0.1 to several mA) with a width of 0.5 to 2 ms is applied to the ventricle at a fixed rate of about 60 pulses per minute through a flexible, catheter-like conducting electrode that runs from pacemaker to ventricle, usually through a vein ("asynchronous pacing"). In a more physiological mode, one electrode affixed to the atrium detects the atrium's depolarization, which is recognizable as the P-wave of the ECG; after an appropriate delay during which the ventricle is filled with blood, the pacemaker delivers a pulse to the ventricle ("atrium-synchronous pacing"). If the heart block is not total, normal conduction may be disrupted only occasionally. In such cases, the pacemaker is programmed into a mode, which applies a pulse to the ventricle only if the ventricle's depolarization does not occur within a certain time after the atrium's depolarization has been detected ("demand pacing").

Knowing whether a patient has an implanted pacemaker is important, because defibrillation may damage or reprogram it. In elective cases, the patient's record will show this information (if it can be found!); in emergencies, records are often unavailable. The ECG will often show whether a pacemaker is present. Sometimes, pacemaker pulses are directly visible in the ECG. Pacing pulses are so short and their energy is so low, however, that they are frequently invisible on the display of standard ECG-monitors, whose bandwidth is limited. Yet, even then a more detailed analysis of the ECG can often show the presence of a pacemaker. In asynchronous pacing, the ECG-derived heart rate, which is based on the time between successive QRS-complexes, is utterly constant, whereas a non-paced ECG always shows some irregularities. In atrium-synchronous pacing, likewise, the time between a P-wave and the following QRS-complex is more constant than normal. In demand pacing, however, the presence of a pacemaker cannot be discovered as long as the patient's ventricle functions normally; then the pacemaker is "silent."

8.6. ELECTROCAUTERY

Besides their lancet, surgeons use an "electric knife." Electrocautery equipment, also referred to as diathermia[1] equipment, the "Bovie" or the electrosurgical unit (ESU), uses a strong high-frequency electric current whose heat production, as the current enters the tissues, can be used for cutting and/or coagulation.

The term *cauterization* refers to destroying tissue through a variety of means: chemical corrosion, heating, freezing, or electrical energy. In electrocautery, it is the current's very local thermal effect that cauterizes (burns) the tissue. In contrast to defibrillation, where the tissue contact resistances are made as small as possible, one of the two contact resistances is now large. The active electrode has a fine tip; various shapes and sizes of tips (probes) are available for specific procedures. In fact, parts of the probe may not even touch the body; the high voltage then causes a plasma arc discharge current. Powers up to 800 W are used at frequencies between 500 kHz and 3 MHz. Not all the device's energy is converted to heat at the site of cauterization; some energy reaches other tissues and organs. In the heart, arrhythmias can arise. Experience has also shown that at lower frequencies nerves and muscles are stimulated. Yet, even at the frequencies used some stimulation is unavoidable, especially in moist tissues such as the bladder. This is due to the fact that some rectification takes place at the electrode-tissue interface. It is the fluctuations in the rectified current that cause the stimulation.

At low power, the voltage is applied pulse-like (Figure 8.15) and coagulation takes place. When tissue coagulates, it becomes more solid, as when an egg is boiled. Coagulation stops bleeding and closes wounds. At high power, the voltage is applied continuously; the tissue first coagulates, then dehydrates (its water content evaporates), and finally it is incinerated and blown away. At intermediate power levels, a blended waveform is used; cutting and coagulation take place simultaneously, which results in "bloodless cutting."

In the *monopolar technique*, only one electrode is active. The second electrode has a large surface area and a low resistance path to the patient's body. If the naked patient lies on an operating table, this "neutral" electrode is the conducting (upper layer of the) pad on which the patient lies. In order to provide a low resistance return path, the pad must be large, it must be moisturized with a well-conducting contact fluid, and it must contact body areas with a good blood supply (e.g., buttocks, thighs), which can satisfactorily disperse the heat generated at this contact site. When the power is very low, as in dentistry, ophthalmology or neurosurgery, there need not be a neutral electrode; the equipment's second pole is grounded and the circuit is closed through the patient's capacitance to ground.

In the *bipolar technique*, both electrodes are active; the two-electrode assembly looks like a tweezer. The high-frequency current now only flows through the tissues between the tweezer. This avoids damage to surrounding tissues and requires less

[1] The term *diathermia* can create confusion, because it has other medical meanings as well.

power. Bipolar electrodes are used mostly to coagulate and close small blood vessels.

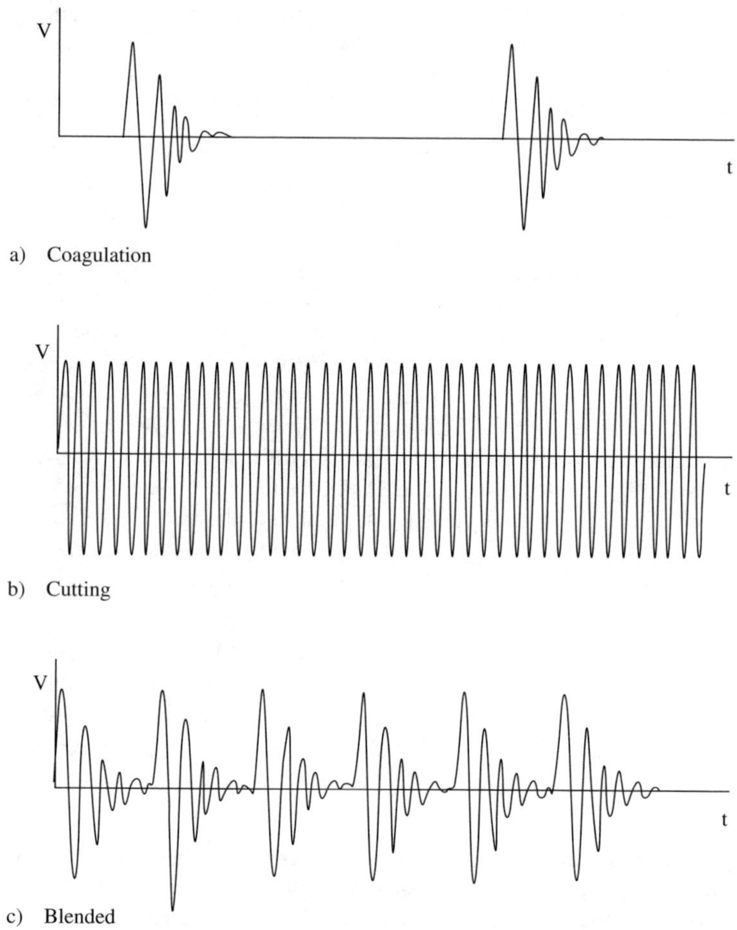

Figure 8.15. Electrocautery voltages for coagulation (a), cutting (b) and blended mode, i.e., cutting plus coagulation (c).

The thermodynamic balance at the electrode-tissue interface determines the effect. Factors that determine the heat flow *to* the tissue are:

- The current density, which is determined by the specific resistance of the tissue and the shape of the electrode. The added heat energy is proportional to the specific tissue resistance and to the square of the current density.
- The duration of energy application. This depends, given a certain electrode shape, on the velocity of the electrode with respect to the tissue.
- The current's duty cycle or power factor: continuous, pulsed or blended.

Factors that determine the heat flow *away from* the tissue are:

- The tissue's heat conductivity.
- The heat conductivity of the tissue's surroundings, especially moisture and liquids.
- The tissue's shape, especially whether and how it is lacerated.

Choosing an appropriate power setting based on these factors is a matter of considerable experience.

Electrocautery units are common devices in the OR. Their use requires skill and care; they have been known to cause shocks (if the ground path's resistance is high, current may discharge through the OR personnel), burns (if the ground path's resistance is concentrated in a small area of the pad), explosions (if combustible anesthesia gases are used), arrhythmias, and pacemaker interference (suppression of stimulation; even reprogramming and destruction can occur). Burns may occur at unlikely sites. If the connection to the neutral electrode is interrupted, burns may occur, for instance, at ECG electrodes, which then provide a return path for the current.

Due to its high energy, an active electrocautery device disturbs many of the measured signals, especially the electric signals (ECG, EMG, EEG). If the surgeon electrocauterizes for long periods, a pause may be requested at intervals, so that the patient's vital sign values can be updated. Research continues to make the monitoring equipment less sensitive to electrocautery.

Closely related to electrosurgery is laser surgery. Here, too, the thermal effect is used. The limited power of lasers restricts their use to micro-operations such as in ophthalmology. Fiber optics conduct the laser beam to the desired location. The (invisible) laser beam is usually accompanied by a beam of visible light, so that the surgeon can see where the beam is pointing, even if the laser itself is not switched on. Laser surgery, too, requires considerable experience.

QUESTIONS

1. Give a model for the infusion flow in an infusion drip system and use it to discuss the factors that determine the flow.
2. Discuss the advantages and disadvantages of ventilating a patient with a ventilator that provides a) a controlled flow, b) a controlled pressure to the patient's airways.
3. What is the function of a ventilator's inlet combination?
4. Describe how the open system ventilator, whose diagram is given in Figure 8.5, functions. When do the various valves open and close, and what determines this?
5. Derive the curves of Figure 8.7 from the diagram of Figure 8.6.
6. What is "assisted ventilation" and how is it realized?
7. Explain why PEEP is required when ventilating a patient with pneumothorax.

8. Describe how the circle system ventilator, whose diagram is given in Figure 8.11, functions. When do the various valves open and close, and what determines this?
9. Some commonly occurring problems in artificial ventilation are disconnection of inspiratory or expiratory hose; leak of inspiratory or expiratory hose; esophageal intubation; bronchial intubation. Which measurements and/or patient data would be required to discover these problems in an alarm system?
10. Draw the schematic diagram of a heart-lung machine. Explain the function of each block.
11. What is a defibrillator? What is a cardioverter?
12. What is electrocautery? How is it realized? What are the different voltage waveforms for?

9 Patient Monitoring

Whether in patient monitoring or elsewhere, physicians have similar concerns. When we analyze how a medical doctor — whatever his or her specialty — uses measurements, we notice several things. First, the doctor strives for certainty. Given the great importance of the patient's health, it is considered a professional failure if an incorrect diagnosis is reached and thus an incorrect therapy is instituted. Thus, the doctor wants to have all relevant information that will enable determination of the appropriate diagnosis. This requires the clinician to have an up-to-date knowledge of what to look for and an up-to-date experience of problems that may arise. It also requires the most informative measurement devices with the greatest accuracies and the lowest error rates. Requirements for measuring devices therefore include guaranteed accuracy: a well-designed instrument will indicate when it cannot present an accurate number, and not present an inaccurate or random number. Instruments should not mislead the clinician. It is often possible to arrive at the correct conclusion when a piece of evidence is missing, but *not* when incorrect information is believed to be accurate. A device must also present its results in such a way that no confusion is possible regarding the numerical values. This calls for good ergonomic design and a certain degree of simplicity in the presentation of results.

Second, the doctor needs to use his time for the patient, not for the instruments. Instruments that take too much time to set up or keep in good operating condition will not be used. Medical device manufacturers are caught between two conflicting design requirements. One is to have a device that is as flexible and as informative as possible, but this may come at the cost of many knobs to turn or many menus to choose from. The other is utter simplicity and reliability of operation. Ideally, the instrument should be extremely easy to set up and require no maintenance while in use (e.g., calibrations or recalibrations, setting changes, cleaning of parts, repair of defects), yet allow all secondary information to be easily accessible during less stressful times. The future is for devices that check themselves for possible measurement problems, that understand from moment to moment what the doctor wants, and that automatically adapt to these requirements. Such a device can be thought of as a robot, an autonomous device that knows what to do and that does it, without having to be told what to do.

Some of these "robot" systems will take the form of control systems that take over time-consuming routine tasks. For instance, if a too high blood pressure is *always* treated with a certain flow rate setting change of an infusion pump that delivers a drug that lowers the blood pressure, this part of the care process can be automated — provided it can be done reliably. If a flow rate setting needs to be changed only *under some conditions*, the process can still be automated if these conditions can be reliably detected by the device.

In the third place, physicians are — like all humans — easily overloaded by too much cognitive information, especially if there is little time to evaluate all that information. This is generally the case in critical care environments such as the operating room (OR) and the intensive care unit (ICU). This observation conflicts with the first point. If there is sufficient time for study, the doctor wants *all* the information. If there is little time, the doctor wants only the *relevant* information. But the decision of what is relevant in a certain situation is a medical one. If a device is to decide what is relevant, it needs to have extensive medical knowledge built in. So-called "intelligent alarm systems" are examples of this approach. Ideally, they would show nothing when everything is as it should be, or just that fact. When a problem develops, they would just show the data that indicate the problem and its most probable cause(s). Such a device can be thought of as an extra "specialist" with a great deal of knowledge about a very restricted knowledge domain (what is relevant to display when), an autonomous robot-like device that knows when to come into action and what to do at that time.

9.1. MEDICAL DECISION MAKING; DIFFERENTIAL DIAGNOSIS

In order to understand what measurements are used for, we need to briefly consider the process by which a physician reaches a diagnosis. Appropriate *action* (therapy) presupposes appropriate *knowledge* (diagnosis) of what the problem is. This knowledge is acquired in a variety of ways, usually as follows:

a. Patients tell the physician what their complaints are and which expectations they have concerning treatment. The patient usually expects healing or at least lessening of complaints, but may impose restrictions on treatment. In addition to what the patient reports, the physician asks a number of questions that provide a more complete picture of the problems. The total information that exists at this point is called the *medical history*.

b. A more or less thorough *physical examination* takes place, which focuses on the patient's complaints.

c. At this point, the physician forms a number of *hypotheses* about the underlying cause or causes of the complaints.

d. Extra investigations (more questions, a more detailed physical examination and often tests in a medical laboratory such as analysis of blood or urine, X-rays or an ECG during exercise) are used to confirm or exclude the hypotheses formed earlier. The goal of these *lab tests* is to confirm one hypothesis and reject all others; this process is called *differential diagnosis*. If the goal is reached, the one remaining hypothesis is the *diagnosis*. If all hypotheses fail, more knowledge has been obtained and the physician can form better hypotheses (back to c).

In some cases such a complete elimination process is unnecessary or even impossible. Sometimes that is no problem; if there is a single therapy that covers all the remaining hypotheses, that therapy can be given. If the problem is, for instance, an infection caused by one of several types of bacteria, a certain antibiotic may be effective against all of them. If uncertainty remains as to the underlying cause and no single best therapy is possible, either the patient is referred to a specialist (who restarts at a, b or c), or the therapy for the most likely problem is started.

e. At this point, the diagnosis has been established; it implies a number of possible therapies. The earlier obtained information (e.g., a drug allergy) can exclude some of the therapies. If no purposeful therapy is available, reduction of the complaints (e.g., pain) is the only remaining option.

f. All remaining possible therapies are now ordered according to desirability, based on factors such as cost, duration of the therapy, probable side effects and how well the patient will tolerate the therapy.

g. At this point, the most promising therapy is chosen.

h. If the chosen therapy does not have the expected result, more knowledge has been obtained and a better diagnosis can be made (back to c), or a different therapy is chosen (back to f), or a different dosage is tried.

This procedure is the same for every physician, regardless of specialty. The diagnostic and therapeutic possibilities and the time that is available to reach a decision, however, may vary a great deal between specialties.

9.2. RISK AND QUALITY OF INFORMATION

Measurement technology is a discipline which, among other things, focuses on the extraction of reliable information from all sorts of measurements. In a medical environment, extra considerations play a major role, related to the vulnerability of the subject from whom measurements are obtained and with the "noisy" environment in which the measurements take place (see, e.g., Section 8.6).

The physician's primary ethic, which is taken very seriously, is *do not harm*. For measurements, this implies a risk-benefit analysis. Measurements that cause more harm than provide useful information are considered unethical, in particular measurements that will never be *useful* for that patient.

Thus, the type of measurement must be adapted to the patient's health condition. In an otherwise healthy young patient who is operated for an appendicitis, a minimum amount of monitoring may be indicated, whereas a multi-injury trauma patient who has barely survived a car accident and is in a critical condition may require the most intensive (and thus risky) monitoring that is available if the patient is to have a chance to survive.

In a choice of measurement method, several options are available. One option is the choice of invasive versus non-invasive measurements. In invasive measurements, the patient's body must be "invaded" through disruption of the skin. Generally, these measurements provide more or better information, but they can also have more serious consequences, such as tissue trauma and the risk of bacterial infection. An example of this choice is in arterial blood pressure measurement: should a cuff method or a catheter-manometer method be used?

Another option is the choice of direct versus indirect methods. Generally, direct methods provide more or better information at a higher rate, but they are also more invasive. An example of this choice is in arterial blood flow or cardiac output measurement: should the electromagnetic blood flowmeter method, a dye dilution method, or Fick's method be used?

A related option is the choice of continuous versus non-continuous measurements. Continuous measurements are usually invasive and direct. Continuous measurements give essentially instantaneous results. The frequency at which non-continuous measurements may be repeated varies. In a critically ill patient, the necessity to have values at least every minute will often exclude non-continuous measurements, often leaving invasive measurements as the sole option.

The most critical decision criterium is the stability of the patient's condition. If the condition does or could start to fluctuate rapidly, continuous or frequent measurements are required. Since in case of a major calamity — standstill of respiration or circulation — the patient may die within 3 min, the measurement frequency should be higher than once per minute to assure that sufficient time remains for action. When a measurement method is to be chosen for a critical care environment such as operating room, intensive care unit, neonatal intensive care unit or coronary care unit, this consideration is the most important one.

The criteria for choosing a measurement method thus depend on patient-related circumstances. But other criteria play a role as well. In medical research, one will generally want the most accurate and informative method. Such methods are usually the most invasive ones. Since these carry more risk for the patient than strictly necessary, medical ethics require that the patient must have given consent for participating in a study.

In routine OR and ICU cases, other concerns play a role. Because the measurement's information must be weighted against the possible harm that measuring may cause, there is a strong tendency toward non-invasive measurements. Here, clinicians are limited in their options: since most equipment is purchased, the manufacturer of integrated systems such as patient monitoring systems, ventilators or heart-lung machines has already made the choice. This choice usually depends on cost, the manufacturer's experience and impression of what the user wants, and the tendency to imitate successful designs of others. Reliability is a major concern. Device failure may result in extremely expensive lawsuits and may ruin a manufacturer's reputation. Thus, designs tend to be conservative ("proven technology"). Because of the importance of the cost factor, devices are often not the most accurate. Moreover, accuracy is often difficult to establish clinically. This is the problem of the missing *gold standard*: if the true value could be accurately measured, one could fruitfully determine how closely a device approaches this ideal. If not, a clinical comparison is often unproductive.

Hospitals' concerns are cost and serviceability, which depend on the technicians' familiarity with devices. A too large range of different devices is not recommended.

Question: An otherwise healthy woman undergoes surgery for appendicitis. She will be artificially ventilated. Monitoring is mainly required to detect human error. Which measurement devices would you suggest?

Question: A car crash victim undergoes major emergency surgery for severe respiratory and circulatory problems. He shows multiple lacerations and massive bleeding and his condition is very unstable. Which measurement devices would you suggest now?

9.3. SAFETY ISSUES

Measurement equipment should never unintentionally harm the patient. Very strict safety norms apply to medical devices. The *first fault condition* specifies that whenever one fault occurs in a device, this fault, *whatever it is*, may not result in danger for the patient (e.g., dangerously large electrical currents). Electrical device safety is only one aspect; leakage currents must be especially small when sensors can come into contact with the patient's skin, and even more so when they can contact the well-conducting interior of the body.

Even if the device is in perfect working condition, safety must be ensured. Besides direct harm for the patient, a measurement device can cause indirect harm if it is employed in such a way that it returns incorrect numbers. Physics tells us that every measurement disturbs the measured object. In many biomedical measurement methods, this disturbance is large. This must always be taken into account, since it limits the applicability of a method. The first safety requirement is, of course, that the measurement does no unintentional harm. The second is that the measurement does not disturb the physiological system whose function is measured. The third is that the measurement should not disturb the quantity to be measured.

We will give an example of each of these requirements from the invasive arterial pressure measurement. First, access to the artery must be provided with as little tissue damage as possible; this is a matter of clinical experience. Second, the cannula that is used in the measurement must not be so large that a major block of the blood flow to an organ results, which could cause damage. Third, with an upstream pressure measurement, the pressure reading may be unrealistically high (see Section 3.2.1.1).

Several safety issues beyond the normal ones exist in a patient-monitoring context. Since the patient's condition can vary rapidly, the continuous availability of measurements must be ensured as well as possible. Many measurements can be disturbed by other equipment or by actions of the surgeon or anesthetist. Movement of the patient can cause small displacements of sensors, resulting in measurement artifacts. These movements can also be caused by surgical manipulations. Electrocautery (see Section 8.6) and external defibrillation (see Section 8.4) can cause large electrical currents and high voltages in and around the patient; devices must be able to withstand these.

Knowledge of the context in which measurements take place is essential for an understanding of disturbances in signals. In an invasive arterial blood pressure measurement, for instance, small air bubbles are sometimes flushed into the bloodstream. During a flush, the pressure reading is disturbed (see Section 3.2.1.1). Larger air bubbles cannot be flushed into the bloodstream, because they might, when they reach an artery so small that it cannot accommodate the air bubble, block the blood flow through vital tissue (such an air bubble is called an embolus). Through manually repositioning a three-way stopcock, large air bubbles are therefore flushed into the outside air (Figure 9.1). The existing access to the bloodstream may also be used to draw a blood sample or to inject a drug. This is done by repositioning the stopcock, so that the syringe has access to the bloodstream. During the duration of blood sampling or injection, the pressure measurement is of course invalid. The measurement is also invalid when the line is flushed afterwards. After drawing blood, this is necessary to return the blood that remains in the cannula — where it will clot — to the artery. After an injection, flushing is necessary to wash the drug that remains in the cannula into the artery.

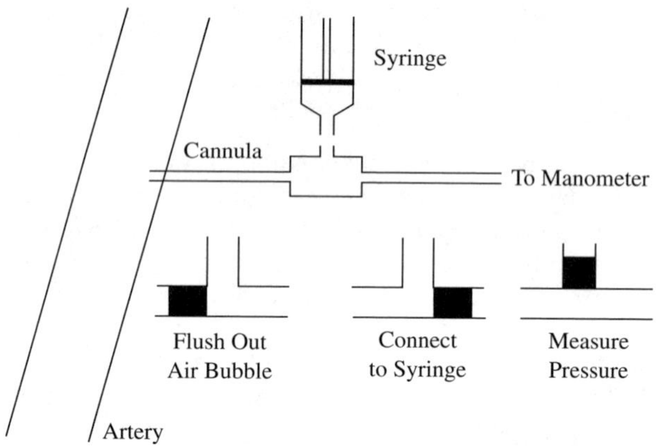

Figure 9.1. Invasive blood pressure measurements may be temporarily (20 s to several min) unavailable due to flushing of the catheter or blood sampling.

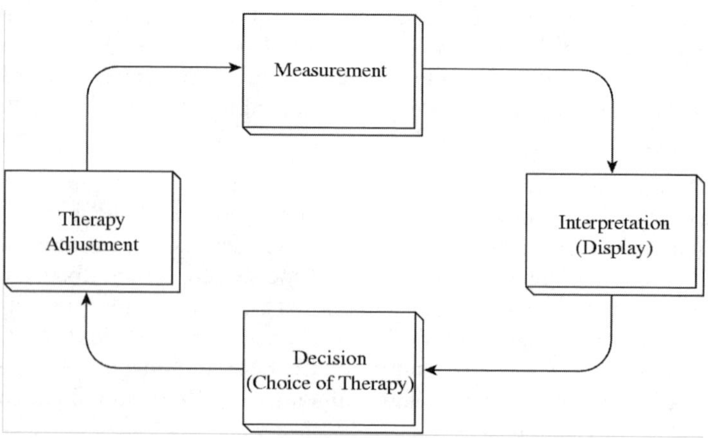

Figure 9.2. The therapeutic feedback loop. Interpretation of the measurements may lead to an adjustment of the therapy or a different therapy.

While valid measurements are absent, the therapeutic feedback loop (Figure 9.2; see also Figure 9.7) is cut. Shannon's sampling theorem tells us that the minimum rate at which the measurement should be repeated depends on the highest frequency in the signal. Therefore, the time during which the signal can be allowed to be absent depends on the stability of the patient's condition. A clinician can often use information that is present in other signals to estimate a time period during which the last valid measurement can still be used. In an instrument that does not have access to other signals (see Figure 9.7), this will not be the case; for reasons of safety, it will need to signal that it cannot continue to control after a relatively short time period, at which moment the clinician should take over. A similar analysis in a general monitoring context (see Figure 9.2) shows that outdated measurements may lead to an incorrect and possibly lethal therapy.

9.4. PATIENT MONITORING AS A PROCESS

Patient monitoring is a systematic process of observation, interpretation and evaluation of the patient's condition with the goal to bring the condition toward or keep it at what the clinician thinks is optimal for the patient. This optimum is usually operationally defined as a prescribed range for certain variables, e.g., "the heart rate should remain between 60 and 80 beats per minute." Thus, patient monitoring is a feedback process, as indicated in Figure 9.2. The figure clearly shows that diagnosis is not a once-only event; it is required continually, because the patient's condition is or could become unstable.

Patient monitoring predominantly takes place in several specialized units of a hospital. In the 1960s, external defibrillation devices (see Section 8.4) became available. They made it possible to save the victims of a sudden heart attack without an emergency operation, which was required before that time to provide access to the heart. If these patients were located throughout the hospital, assistance often came too late. More success came about when all patients for whom sudden heart standstill threatened were concentrated in one unit where the necessary equipment and trained personnel were constantly available. Thus, the *coronary care unit* (CCU) came into existence. The first CCUs were designed for rapid defibrillation in event of a heart attack and mainly relied on ECG monitoring. Soon it was discovered that heart standstill is often preceded by arrhythmias (heart rhythm irregularities), and it was also discovered that therapies that prevented arrhythmias also prevented heart attacks. As soon as the types of arrhythmias that precede standstill were identified and appropriate therapies were designed, the emphasis in the CCU shifted to prevention. Monitoring became less critical and was concentrated in one central station, where the ECGs of all patients are presented and where automatically generated alarms sound when rhythm irregularities are discovered.

The *intensive care unit* (ICU) is meant for critically ill patients who require more care than is available elsewhere. These patients form a more heterogeneous group than those in the CCU and thus require a larger variety of measurement devices. In large hospitals, more specialized intensive care units exist, where patients are more homogeneous and where monitoring devices thus can be better standardized. In ICU patients, the respiratory and circulatory systems must, due to their time-critical nature, be monitored most frequently and intensively.

The *operating room* (OR) is the domain of surgeon and anesthetist. It is the anesthetist's task to keep the patient in such a condition that the patient can successfully undergo an operation by the surgeon. In major operations, this condition implies painlessness, unconsciousness, amnesia and muscle relaxation. The last is required in order to prevent reflex movements, but it also means that the patient's respiratory muscles do not function anymore and that the patient must be artificially ventilated. Elimination of the patient's consciousness and reactions to pain require an "external intelligence," that of the anesthetist, to function as a "patient monitor," usually supported by a large variety of measurement equipment.

9.5. NORMALITY AND STABILITY

Measurements play a primary role in establishing the diagnosis. Two concepts play a central role in the interpretation of the measurements, *normality* and *stability*. An organ, an organ system or a physiological function is *normal* if it functions in an adequate way, and abnormal if its function is so disturbed that the organism does or could experience trauma (damage). What is normal depends on a large number of personal factors such as the patient's age, weight, constitution (thick or thin), etc. Whether a measurement is normal can often be evaluated only in relation to other information.

Whether a measurement is normal may also depend on the time when it was acquired. If the patient's illness is more or less *chronic* (unchanging), one measurement may be representative for a long period. Sometimes we see a *trend* (a slow change), a worsening of the problem or a spontaneous remission. Sometimes the problem is *acute* (occurs suddenly) or *variable* (sometimes severe, sometimes almost or completely absent). In such cases it is important to know how *stable* the function is. A heart infarction, for instance, may occur quite suddenly. Measurements must be performed very frequently or even continuously if a problem could appear suddenly and can have serious consequences.

9.6. WHICH PHYSIOLOGICAL FUNCTIONS TO MONITOR

In order to decide on what to monitor in a specific patient, it is important to know what can be monitored. The following list is far from complete; new devices appear continually. Many devices, however, do not live up to expectations. They may be less accurate than practice requires, or demand more of the clinician's time than the extra information that they offer warrants. The following measurements can be performed with commonly available devices:

Respiratory: airway pressure and derived variables; airway flow and derived variables; gas concentrations of CO_2, O_2, N_2, N_2O, several anesthetic agents and derived variables.

Circulatory: electrocardiogram (ECG) and derived variables; blood gases, blood electrolytes and variables derived from blood properties, e.g., hematocrit and thromboelastogram (TEG); transesophageal echocardiogram (TEE); non-invasive arterial blood pressure; invasive arterial, central venous, and pulmonary artery and wedge pressures; oxygen saturation (SaO_2); cardiac output (CO).

Other: electroencephalogram (EEG) and evoked potentials (EP), electromyogram (EMG), degree of muscle relaxation, core and skin temperatures.

Maxima, minima, means, periods or frequencies and other important *derived variables* can be extracted from the (quasi-)periodic signals. Examples are systolic arterial pressure, minimum expiratory pressure, mean arterial pressure, heart rate, respiration rate, ratio of inspiration and expiration time (I/E ratio). Simple (e.g., integration, differentiation) or more complex algorithms can provide extra features such as tidal volume and minute volume (respiratory flow), presence of arrhythmias and ST-segment elevation (ECG) and arterial dP/dt (invasive arterial pressure).

On the other hand, one can specify a *minimum set* of measurements to be obtained or monitors to be used in, e.g., an operating room. Many countries enforce such norms, which differ from country to country but show an emerging consensus. During general surgery, for instance, generally accepted minimum standards include the end-tidal CO_2 concentration ($ETCO_2$), the electrocardiogram (ECG), the arterial oxygen saturation by pulse oximetry (S_{pO2}) and the noninvasive arterial blood pressure (NIBP), with the ability to measure invasive blood pressure, and temperature. The American Society of Anesthesiologists (ASA) was one of the professional societies that formulate minimum objectives and how these can be best fulfilled. Their 1986 norms were:

Oxygenation: to ensure adequate oxygen concentration in the inspired gas and in the blood. Measure the oxygen concentration in the breathing system; include a low limit alarm. Visual access to the patient must allow assessment of skin color. Pulse oximetry is encouraged.

Ventilation: to ensure adequate ventilation. Use at least qualitative signs such as chest or bellows excursions. CO_2 monitoring is encouraged. Detection of disconnections in the breathing system is required, including a disconnect alarm.

Circulation: to ensure adequacy of the circulatory functions. Required are continuous ECG; arterial blood pressure and heart rate at least every 5 min; at least one of pulse palpation, heart sound auscultation, intra-arterial pressure monitoring, ultrasound peripheral pulse monitoring, pulse plethysmography or oximetry.

Body temperature: to aid in the maintenance of appropriate body temperature. Continuous temperature measurement is required.

9.7. MONITORING DEVICES

Monitoring devices come in varying degrees of complexity, from a simple temperature sensor or manometer to very complex systems (e.g., the Peñáz method). Basic requirements of transducers are:

- faithful reproduction of the dynamic range and the frequency range of the physiological variable;
- reproducible manufacturing process (if for one-time use) or easy to recalibrate (if reusable);
- linear or linearizable;
- simple to apply, maintain and remove;
- not susceptible to external influences (movement, electrocautery, etc.);
- harmless in terms of electrical current, mechanical stress and infection;
- inexpensive (if for one-time use) or easy to sterilize (if reusable).

What all modern measurement devices have in common is that they provide an electric output signal that can be easily processed and displayed.

9.7.1. Signal characteristics; bandwidths

In environments where patient monitoring takes place, the respiration frequency is normally lower than 30 breaths per minute (0.5 Hz) and the heart rate less than 120 beats per minute (2 Hz). Faithful reproduction of these signals requires a bandwidth that passes up to 20 (in critical applications up to 50) harmonics without distortion. For respiratory signals, this implies a bandwidth of 10 to 25 Hz, and for circulatory signals a bandwidth of 40 to 100 Hz. This poses few problems for the electronics in a device; it is usually the transducer that imposes bandwidth restrictions and may have an under- or overdamped transfer function.

9.7.2. Signal acquisition, validation and processing

Although the problem *which* data should be acquired and analyzed is application-dependent, in modern integrated monitoring systems the basic signal processing procedures will often be comparable. Algorithms perform the acquisition and processing of a more or less standard set of similar measurements and extract similar features: maxima, minima, periods, slopes, etc.

Two problems have a general character. The first is that the acquisition rate of many of the measurements is so high that no human would be able to handle the "raw" data; some sort of data preprocessing is required so that only more meaningful data will be offered at a much lower rate. The second problem is that the quality of the data is to be suspected. Due to a variety of causes, the acquired data may not reflect the quantity that they are supposed to represent. A process of *data validation* is required to establish the authenticity of the acquired data. Signal processing must isolate the clinically significant features of the signal; data validation may be based on some additional features as well, such as consistency between data items. A body weight of 80 kg combined with a body length of 60 cm should raise suspicions and, if possible, prompt for correction. The latter is possible only for manual data entry, however.

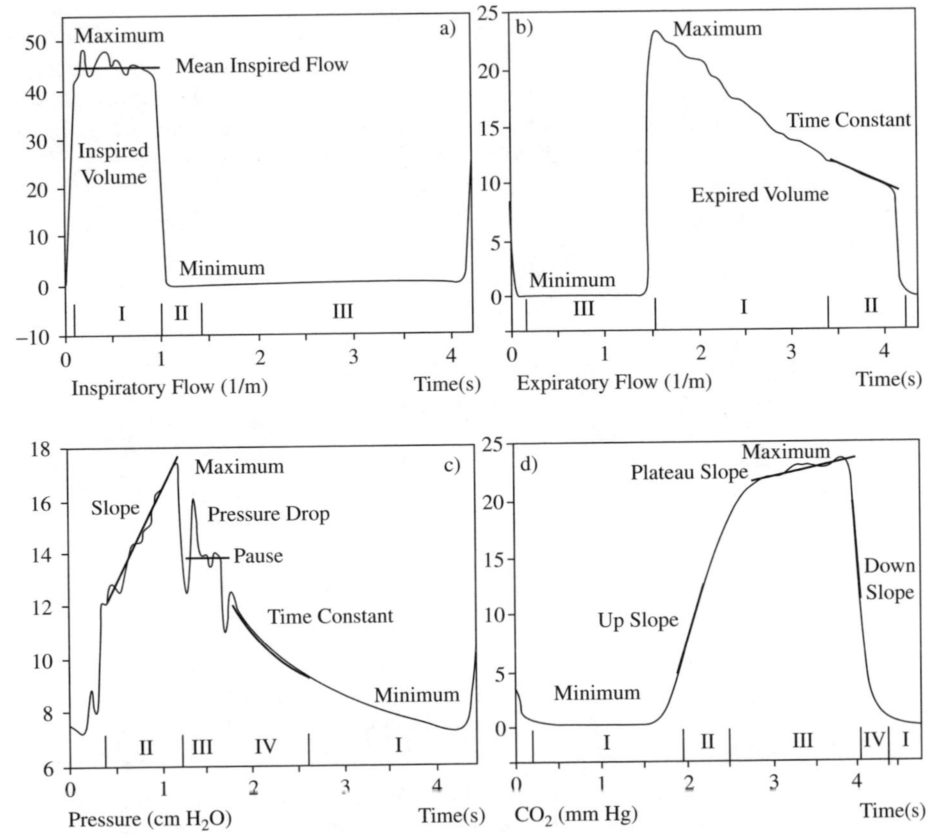

Figure 9.3. Measured inspiratory (a) and expiratory (b) flow, respiratory pressure (c) and CO_2 concentration (d) signals frequently deviate from the "ideal" signals generated by a model. Dog experiment.

Figure 9.3 shows an example of some actual "raw" respiratory signals measured in a dog (the shapes of these curves were explained in Chapter 6 and in Section 8.2). The curves are far less "clean" than the idealized curves that have been presented thus far. Yet, the idealized curves are guides that tell us which features should be extracted. The inspiratory flow signal must yield the maximum, minimum and mean flows, the inspired volume and the inspiration period. The expiratory flow signal must similarly yield the maximum, minimum and mean flows, the expired volume and the expiration period; in addition, an exponential is fitted to the decreasing part of the curve in order to obtain the "time constant," the product of expiratory resistance and lung-thorax compliance. The pressure signal must produce the maximum, minimum and mean pressures, as well as the stepwise initial pressure increase, the inspiratory slope, the inspiratory pause pressure, and the time constant of the expiratory part of the curve. The latter parameters allow some of the properties of the respiratory circuit to be estimated. From the CO_2 signal, minimum and maximum concentrations as well as several slopes must be determined. Data processing algorithms often need to perform simple operations only: discover discrete points (maximum, minimum, starts and ends of slopes) in the curve or in its low pass filtered version, fit linear or exponential functions to signal sections (slopes, time constants), or determine averages (means). Some algorithms need to perform much more complex operations, such as the pattern recognition that is required to discover abnormal QRS-complexes in an ECG signal.

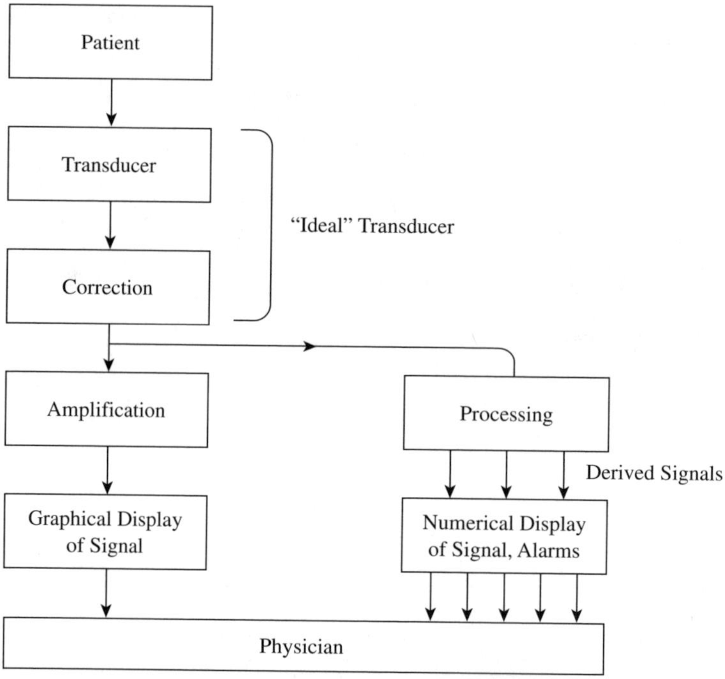

Figure 9.4. Data processing steps. A signal is acquired by a transducer whose characteristics may need correction. After "dumb" but faithful amplification, the signal is presented as a graph. After possibly very complex signal processing, the signal's features are presented as a number or a set of numbers.

Figure 9.4 shows a schematic of a classical single signal monitor. A transducer acquires the signal. Because no transducer is perfect and exactly reproducible, some type of correction may be required, e.g., for non-linearities or fluctuations in characteristics from device to device.[1] In an invasive arterial pressure measurement, for instance, a calibration sequence may be needed in order to generate an "ideal" signal that is independent of an individual transducer's offset and gain. The offset calibration can be combined with the "null" calibration, which is required anyway to eliminate the hydrostatic pressure component (see Figure 3.5) after the manometer

[1] In modern transducers, this correction has already been performed as part of the fabrication process, e.g., through "laser trimming" of components.

has been put into position. Part of the correction may be filtering to eliminate noise and unwanted signal frequencies.

After correction, the signal goes two ways. First, it is visually presented onto an oscilloscope or monitor display, because the signal's shape contains information about measurement- and patient-related problems. Second, the signal is processed and its derived features (maximum, minimum, period, etc.) are presented as alphanumerics. The values of the derived features are also compared with alarm limits; when these are exceeded, a visual and/or auditory alarm is generated.

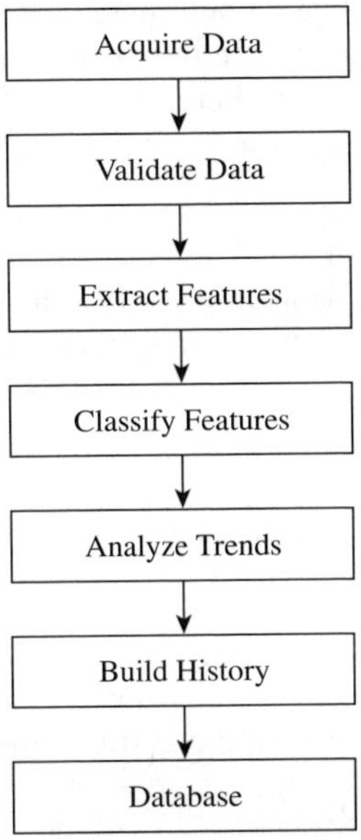

Figure 9.5. Data processing steps in a modern patient data acquisition system.

Figure 9.5 provides a conceptual model for the processing of data in a modern, computer-based patient monitoring system. The data are acquired by the basic measurement devices, as depicted in the top part of Figure 9.3. The data validation process determines whether the data are valid or artifactual. If the data contain considerable redundant information, e.g., in case of a waveform, the feature extraction algorithms extract all meaningful features from each period of the signal, as in the rightmost part of Figure 9.3. If one or more of the extracted features are abnormal, the feature classification process attempts to determine the type of abnormality. The trend analysis process classifies and analyzes the dynamics of the data. The history extraction process builds a compact history of the feature over, e.g., the last few hours.

The problems that must be solved determine the information that must be available in the database. This information in turn determines which signals must be acquired and which features must be extracted from these signals. The characteristics of the signals in turn determine how the features must be extracted.

9.7.3. Data types

Current integrated patient monitoring systems have two major functions. The first is record keeping. Record keeping serves several purposes. During surgery or a patient's stay in an intensive care unit, for instance, the record allows the clinician to review all relevant data. After the case, it allows for statistical analysis and research into factors that determine outcome. Record keeping is also increasingly important in lawsuits, where the records can show that the clinician did all that was possible. Their second function is to realize an automated on-line support system that detects and announces sudden problems before they can cause harm, even if the clinician's attention is temporarily focused elsewhere. For both functions, it is important that all relevant information is entered into the system.

There are four different categories of data that need to be input to such a system: continuous measurements (to be monitored at all times), discontinuous measurements (which provide data only once in a while), data volunteered by the medical staff (also infrequently) and demographic data (which need to be entered only once).

Demographic data. Initially, some important non-changing data about the patient are often entered into the system: the patient's age, height, weight, known allergies, current diseases, etc. A modern approach is to acquire such data from a "hospital information system."

Volunteered data. During an operation, there frequently is important information that, due to limited functionality of the equipment, cannot be acquired automatically but which the staff wishes to impart to the system nonetheless, such as:

- Injections (drug type, dosage).
- Infusions (drug or fluid type, infusion flow rate).
- Fluid loss (blood, urine).
- Settings of ventilator and other equipment that has no electric outputs, such as gas composition, minute volume, anesthetic agent concentration.
- Interventions, clinician-induced occurrences of some event. Many interventions lead to changes or artifacts in signals.
- Artifacts. An artifact is a (sometimes willful and necessary) disappearance or disturbance of one of the signals, due to blood sampling, flushing of a catheter, electrocautery, cardiac output determination, transducer disconnection, power line hum, etc.

Measurement values of this class normally bear a time stamp.

A practical problem with volunteered data is that, due to the pressures of the task, they are not always entered into the system or not entered at the proper time. Volunteered data are useful to an "intelligent" system only if they are coded and can be *interpreted* by the system. If not, volunteered data can only be stored as a "comment" that can be entered into the patient's report.

Discontinuous measurements are usually initiated by the medical staff and performed only if required. Examples are blood gas and/or blood electrolyte determination, cardiac output determination by a thermodilution or dye dilution technique and manual blood pressure measurements using an inflatable cuff. A discontinuous measurement may not be a good representation of the present. Old data represent the patient's past, not the present. Each measurement value of this class should therefore bear a time stamp.

Continuous measurements can often be temporarily disturbed or invalidated, e.g., by movements, actions of the medical staff or electrocautery (see Section 8.6). To obtain a reliable and correct diagnosis based on (features derived from) such signals, the medical staff must be sure that they are valid (not artifactual). This is one of the reasons the most important physiological signals, such as ECG and arterial pressure, are continuously displayed on monitors.

There are periods when the patient's signals are not available: before the transducers are connected, after they are disconnected, during calibrations, etc. During these times no features can be extracted, at least not ones with physiological meaning. A few monitoring devices indicate whether the data that they provide are valid or not; an example is a pulse-oximeter which internally supervises the quality of the signal from which it derives its saturation data. A few other devices indicate that the data that they provide can be or are invalid; an example is an ECG-monitor which continuously measures the electrode impedance: a too high impedance leads to an invalid signal, a low impedance not necessarily to a valid one.

Some of the therapeutic equipment may also be able to make its status known to an integrated monitoring system. A modern ventilator may have outputs for tidal or minute volume, respiration frequency or even complete flow and pressure curves. From these, several features can be computed. Also, setting changes (e.g., a change in tidal volume) can easily be reconstructed from the signals if the equipment does not directly offer such data. In contrast with patient features, equipment features usually do not exhibit spontaneous fluctuations or drift. They are set (and modified) by the medical staff and remain constant over relatively long periods.

The central problem in data processing is the high data rate; new data frequently arrive hundreds of times per second. No human is able to analyze the data at such a rate, but neither is this necessary. Conclusions are usually not based on these "raw" primary data but on features extracted from them. Extracting the features from the data is usually an operation that is performed by *algorithms*. Feature extraction reduces the *quantity* of information to be analyzed by eliminating redundancies in the data and/or irrelevant information; it also reduces the *rate* at which the information must be analyzed.

9.7.4. Display

The system's display is an extremely important link in the total chain from transducer to clinician. Information that is not observed is worthless. The design of the display determines whether and how rapidly information can be acquired by the clinician. This is especially important when vast amounts of diverse, rapidly changing information must be transferred. Vital information must be presented centrally in the visual field. All information that is consulted frequently must be presented within a fairly small visual angle. For these reasons, there is a growing tendency to present *all* the information on one display screen. This display, in its central location, is then separated in space from the transducers, which need to remain in, on or around the patient.

All modern monitoring systems use a computer to acquire and process the data, and a high-resolution computer display to present the data as graphics, text and numbers. The display serves several functions. The most important function is, of course, to present the data that indicate the patient's current condition.

Another important function is signal validation. A real-time display of the most important physiological signals is provided for two reasons. First, the waveform is an indicator for the quality of the measurement. If a low-grade measurement is detected, the derived parameters are suspect and a measurement problem exists. Second, signal waveforms can provide additional information which is not presented as numerical values. Signal validation can partly be automated, but this is not easy: a computer program must be given the knowledge which curves are acceptable and which not. The usual approach is to somehow model the signal as a set of "features" (e.g., minimum, maximum, up- and down-going slopes) that must be between physiologically acceptable limits, combined with the knowledge that (the "features" of) successive periods of the signal must resemble each other.

Another function is that of memory aid; because we cannot expect the clinician to have a perfect memory, the past course of signals or events (e.g., injections, ventilator setting changes) is often important in the interpretation of the present. Displays therefore normally include a number of *trend curves* of derived variables which show the recent past (e.g., 15 or 30 min or 1, 2, 4 or 8 h) of the most important vital signs. These curves are usually annotated by events that were entered into the system and that took place during the same period of time.

Besides curves, numerical data are presented. Unexpected problems arise when we consider *how* numbers should be presented. *Which* information should be presented is determined by the clinician's vocabulary: because anesthetists talk about a systolic arterial pressure, they need to be presented with its value. Mathematically, however, the systolic arterial pressure is a "point process": its value is realized only once per heart cycle during an infinitesimally short time. On the display, we could present a new number each time a new value becomes available. Since systolic values slightly vary from beat to beat even in the most stable patient, this would make it difficult to read the number, especially at high heart rates. Moreover, for record-keeping purposes the anesthetist wants a number that is representative of a longer period. A solution would be to use a low pass filter, but this would be misleading in case of sudden changes. These provide important information, which should not be filtered out. Some compromise is thus required, and, since definitive "best" solutions are unknown, each manufacturer provides a different one.

There is usually far more information than can be presented on a single display. Besides the "raw" signals and the derived data, it may be necessary to present extra information in the form of *relationships* between variables, e.g., PV-loops (see, e.g., Figure 1.29). Different manufacturers choose different solutions. Usually, a default "master" display presents the most frequently needed information, and extra screens can be selected from, e.g., a menu or through dedicated function keys. It is a challenge to find a good compromise between ease of operation and accessibility of the necessary information.

Practical guidelines for display design are

- present only relevant information, do not overload the user;
- present the information centrally and make sure that users can see the display from the locations where they work;
- use familiar display layouts, if possible resembling form layouts to which the clinicians are accustomed;
- label all graphics with names and units;
- clearly relate trends to the time scale and to the numerical information that is derived from the signals;
- present the reliability of the signals and the numbers derived from them.

9.7.5. Modular versus integrated; basic functions

Older monitors were usually designed to measure one single variable. Although with the growing number of measured variables the number of monitors in use became rather large, the advantage of this approach is that only the devices that are actually needed in a certain case are put into operation. There is also a disadvantage: the many different displays, controls and connections in many different locations make it difficult for the clinician to integrate all the presented information and to adjust settings.

Therefore, manufacturers started to integrate several of the most important measurements into a single monitoring system. Some manufacturers designed modular systems. These systems consist of a "box" into which "modules" can be plugged; each module measures a specific variable. Thus, the monitoring system can be flexibly tailored to meet the requirements for each individual case. Most of these systems have a central display and control panel. A problem with these systems arises in complex cases, when more modules are needed than the box has room for.

Integrated systems provide all the monitors that are normally used in one system. Their extra functionality is that they integrate the information from multiple sources.

A heart rate, for instance, can be obtained from several devices (ECG, invasive arterial pressure, pulse oximeter), but there is no need to present three heart rates. Additional advantages are that there is one central display, which presents the data in a uniform manner, and one central control panel, which allows uniform control of all devices. Fully integrated systems do not exist yet; no single manufacturer produces all the devices that are needed in, e.g., an operating room or intensive care unit. Integration of a few extra stand-alone devices into an integrated system is usually difficult or impossible.

9.8. COMBINING MEASUREMENTS INTO A DIAGNOSIS

Traditionally, reaching a diagnosis has been the clinician's task. In patient monitoring, however, nurses play a very important role. With regard to the ICU it has been said that "the nurse is the monitor." Even though the physician still has the ultimate responsibility, the physician has to step in only when the nurse's possibilities are exhausted.[1]

Because of the bulk of data that is generated by all sorts of modern monitoring devices, reaching a diagnosis has become very time demanding. But the time for evaluation is limited and other tasks — e.g., caring for the patient — are also important. For that reason, parts of the diagnostic process have been automated in the form of alarm systems.

9.8.1. Alarm systems

Alarms are intended to draw attention to a problem before it becomes harmful. Most stand-alone monitors can generate an alarm when a monitored variable crosses fixed but adjustable limits. We will call such alarms *static alarms*. These static alarms have several problems:

- They carry very limited information. An alarm may sound if the heart rate goes outside a prescribed range, but it is left up to the clinician to discover the underlying cause.
- An abnormal heart rate may also be caused by measurement problems (e.g., sensor disconnections, calibration errors, electrical noise) or artifacts (due to, e.g., movements of the patient or interventions of the surgeon). The alarm cannot distinguish between these and real, patient-related problems. Because of the large percentage of false alarms (up to 90%), the clinician often disables the alarms altogether.

[1] This creates extra demands for the design of medical devices. They will mainly be used (controlled, interpreted) by nurses, but occasionally also by physicians, who may have different requirements.

- Alarm limits have to be set. Because of the high percentage of false alarms, these limits are usually set very wide, causing true alarms to be given only when a truly critical situation exists. Moreover, if many monitors are used, many alarm limits would ideally have to be set according to the patient's needs. Since this would take quite some time, it is often not done and the built-in default limits are used. These, too, are normally quite wide.
- A very significant change in the patient's condition may occur, which nevertheless leaves the value between the alarm limits. Such changes go unnoticed.
- An unnoticed trend may have already existed for quite a long time before the alarm is finally given.
- Significant heart rate changes can also be caused by physicians themselves, for instance, when they give an injection of a drug one of whose effects is a change in heart rate. Such events are obviously known and should not cause an alarm.
- Too many alarms exist. In a critical care unit, it may happen that 30 or 35 variables are monitored. In case of an emergency, many of the alarms may go off, which only creates confusion — and time delay, because practice shows that silencing of the alarms may have a higher priority than fixing the problem.

Artifacts should not cause alarms. Frequently, however, they are the major source of (false!) alarms. Different solutions have been invented. *Dynamic alarms* can be given when rapid significant changes occur, even within the static alarm limits, and *trend alarms* can be given on discovery of a (slow) long-term trend, which might predict the future crossing of a static alarm limit (Figure 9.6). These extra alarms, however, also show high false alarm rates.

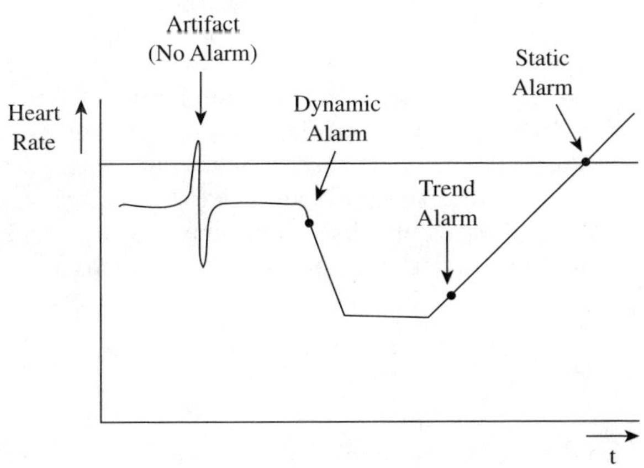

Figure 9.6. Different types of alarms can be given. The straight horizontal lines are "static alarm" limits. Artifacts, usually brief excursions beyond the static alarm limits, should not result in a false alarm.

Several important advances have been made. One is to use the information present in the signals themselves to automatically set appropriate alarm limits. Another is to design better signal processing algorithms that can discriminate between measurement problems and patient-related problems. A third integrates the information present in all the signals and all other entered data into "intelligent alarms" that can be very specific in detecting and alarming the underlying cause of the problem. Redundancies can be usefully employed; the heart rate, for instance, can be crosschecked because it may be obtained from two to four independent sources. Such an alarm system could be an expert system, which incorporates an extensive knowledge base regarding relationships between signals and patient- and equipment-related problems. Such alarm systems can give alarms as specific as "there is a large leak in the inspiratory hose between ventilator and patient" with an incorrect message percentage of less than 10%.

An alarm system can be made even more "intelligent" if it is provided with the ongoing therapy. If, for instance, it knows that an infusion of a drug, which is intended to lower the blood pressure, has just been started, it will not alarm when a blood pressure decrease is detected. Rather, it will alarm when the blood pressure does *not* decrease.

When designing an alarm system, two phases are encountered. In a *top-down* phase, an analysis and inventory of the specific problems to be detected is required. Such an analysis is based on the difficulties that clinicians face and where support is needed. A *bottom-up* phase must then establish how the presence of a problem can be expressed in terms of measured variables. In this design process, it will become clear that certain problems cannot be solved given the available measurements or, alternatively, that certain problems can only be solved if more measurements are made available. Having more measurement devices, however, introduces its own problems. Devices can fail and return erroneous measurements. These must be discriminated from true patient problems. Clinicians need to monitor the patient, *not* the equipment.

The main problem of current integrated systems is their lack of full integration. This is partly due to the fact that no single manufacturer offers all necessary devices. Another reason is that manufacturers are insufficiently aware of the functionality that is required, because clinicians are unable to clearly specify what they require and because clinicians' working styles diverge widely. As a result, current systems display far more data than necessary, have too many displays, and are not able to rapidly provide alarms about underlying causes when problems arise.

9.9. CLINICAL CONTROL SYSTEMS

One of the goals of taking measurements is to check the adequacy of the ongoing therapy. That is also the goal of an alarm system: it warns when problems arise, that is, when measurements deviate from their "normal," nominal values. When a certain problem can be detected with 100% reliability, and when the way in which that problem is to be solved can be specified unambiguously and executed with 100% reliability as well, it is also possible to solve the problem automatically — or better yet, to prevent its occurrence in the first place.

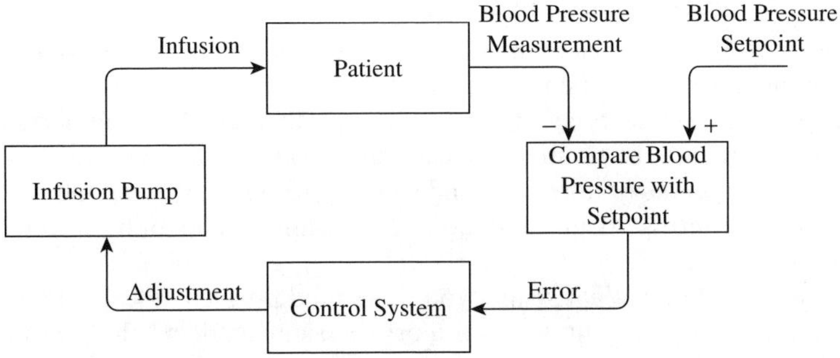

Figure 9.7. A clinical control system assists in the therapeutic control loop.

Clinical control systems take over part of the diagnosis *and of the therapy*. A control system is an extension of an alarm system in that it can operate one or more actuators that perform a certain action (i.e., to infuse a hypotensive — blood pressure lowering — drug). It has:

- one or more sensors that can measure the results of the actions (e.g., the mean arterial pressure);
- a goal, expressed as a desired value for the controlled variable (e.g., a mean arterial pressure of 70 mm Hg);
- a mechanism that relates the measurements and their desired values to actions that must be performed (e.g., a computer program);
- one or more actuators that execute the actions to be performed (e.g., an infusion pump);
- a number of safety mechanisms that detect when control has become impossible (e.g., when the feedback signal is unreliable or when the infusion bag is empty).

An example is shown in Figure 9.7, which can be compared to Figure 9.2. It must be stressed that such a system can only be put into operation when it operates safely *under all circumstances* and shows a performance that at least equals that of the clinician. Safety is paramount; the control system will often have to operate for extended periods (longer than the 3 min that separate a calamity from the patient's death) without supervision. The introduction of these "automatic pilot"-like systems into clinical practice therefore proceeds very cautiously.

QUESTIONS

1. Discuss the conflict in instrument design requirements that medical measurement device manufacturers face.
2. Discuss the conflict in information presentation requirements by medical measurement devices.
3. Describe the process through which a clinician reaches a diagnosis.
4. Give an example how measurement methods' perceived risk and utility may influence the choice of a method.

5. What is the "first fault condition" doctrine?
6. What does Shannon's sampling theorem tell us about the frequency at which measurements ought to be performed?
7. Briefly describe the type of measurements that are most important in a) a coronary care unit, b) an intensive care unit, c) an operating room.
8. Discuss the notions of normality and stability for physiological measurements.
9. What do the terms "data validation" and "signal validation" mean? Research some approaches.
10. Discuss Figure 9.3. What shape would the model-based "ideal curves" of inspiratory and expiratory flow, respiratory pressure and exhaled carbon dioxide have?
11. Discuss the data processing steps of a typical medical patient monitor (Figure 9.4). What is the function of each step?
12. Discuss the types of data that an integrated patient monitoring system must be able to acquire and store.
13. Compare the approximate data storage requirements for systems that store 20 physiological measurements sampled at 200 Hz. Assume the following storage strategies: a) the samples themselves, b) 80 derived variables reduced to a rate of 1 Hz, c) 80 derived variables stored once every 30 s.
14. What are some practical guidelines for the design of a visual display layout?
15. What is meant by the terms "static alarm," "dynamic alarm," "trend alarm," and "intelligent alarm"?

Index